Standalone Renewable Energy Systems

Standalone Renewable Energy Systems—Modeling and Controlling

Special Issue Editors

Rodolfo Dufo-López
José L. Bernal-Agustín

MDPI • Basel • Beijing • Wuhan • Barcelona • Belgrade

Special Issue Editors
Rodolfo Dufo-López José L. Bernal-Agustín
University of Zaragoza University of Zaragoza
Spain Spain

Editorial Office
MDPI
St. Alban-Anlage 66
4052 Basel, Switzerland

This is a reprint of articles from the Special Issue published online in the open access journal *Applied Sciences* (ISSN 2076-3417) from 2019 to 2020 (available at: https://www.mdpi.com/journal/applsci/special_issues/Standalone_Renewable_Energy_Systems).

For citation purposes, cite each article independently as indicated on the article page online and as indicated below:

LastName, A.A.; LastName, B.B.; LastName, C.C. Article Title. *Journal Name* **Year**, *Article Number*, Page Range.

ISBN 978-3-03936-184-7 (Pbk)
ISBN 978-3-03936-185-4 (PDF)

© 2020 by the authors. Articles in this book are Open Access and distributed under the Creative Commons Attribution (CC BY) license, which allows users to download, copy and build upon published articles, as long as the author and publisher are properly credited, which ensures maximum dissemination and a wider impact of our publications.

The book as a whole is distributed by MDPI under the terms and conditions of the Creative Commons license CC BY-NC-ND.

Contents

About the Special Issue Editors vii

Rodolfo Dufo-López and José L. Bernal-Agustín
Special Issue on Standalone Renewable Energy System: Modeling and Controlling
Reprinted from: *Appl. Sci.* **2020**, *10*, 2068, doi:10.3390/app10062068 1

Juan M. Lujano-Rojas, José M. Yusta, Jesús Sergio Artal-Sevil and José Antonio Domínguez-Navarro
Day-Ahead Optimal Battery Operation in Islanded Hybrid Energy Systems and Its Impact on Greenhouse Gas Emissions
Reprinted from: *Appl. Sci.* **2019**, *9*, 5221, doi:10.3390/app9235221 3

Mark Kipngetich Kiptoo, Oludamilare Bode Adewuyi, Mohamed Elsayed Lotfy, Tomonobu Senjyu, Paras Mandal and Mamdouh Abdel-Akher
Multi-Objective Optimal Capacity Planning for 100% Renewable Energy-Based Microgrid Incorporating Cost of Demand-Side Flexibility Management
Reprinted from: *Appl. Sci.* **2019**, *9*, 3855, doi:10.3390/app9183855 33

Eunil Park, Sang Jib Kwon and Angel P. del Pobil
Can Large Educational Institutes Become Free from Grid Systems? Determination of Hybrid Renewable Energy Systems in Thailand
Reprinted from: *Appl. Sci.* **2019**, *9*, 2319, doi:10.3390/app9112319 56

Yinke Dou, Guangyu Zuo, Xiaomin Chang and Yan Chen
A Study of a Standalone Renewable Energy System of the Chinese Zhongshan Station in Antarctica
Reprinted from: *Appl. Sci.* **2019**, *9*, 1968, doi:10.3390/app9101968 68

Yimy E. García Vera, Rodolfo Dufo-López and José L. Bernal-Agustín
Energy Management in Microgrids with Renewable Energy Sources: A Literature Review
Reprinted from: *Appl. Sci.* **2019**, *9*, 3854, doi:10.3390/app9183854 95

Eugenio Salgado-Plasencia, Roberto V. Carrillo-Serrano and Manuel Toledano-Ayala
Development of a DSP Microcontroller-Based Fuzzy Logic Controller for Heliostat Orientation Control
Reprinted from: *Appl. Sci.* **2020**, *10*, 1598, doi:10.3390/app10051598 123

Jian Chen, Bo Yang, Wenyong Duan, Hongchun Shu, Na An, Libing Chen and Tao Yu
Adaptive Pitch Control of Variable-Pitch PMSG Based Wind Turbine
Reprinted from: *Appl. Sci.* **2019**, *9*, 4109, doi:10.3390/app9194109 143

Min Dong, Dong Lv, Chen Yang, Shi Li, Qi Fang, Bo Yang and Xiaoshun Zhang
Global Maximum Power Point Tracking of PV Systems under Partial Shading Condition: A Transfer Reinforcement Learning Approach
Reprinted from: *Appl. Sci.* **2019**, *9*, 2769, doi:10.3390/app9132769 163

About the Special Issue Editors

Rodolfo Dufo-López Electrical Engineer, Ph.D., is Associate Professor of the University of Zaragoza in the Department of Electrical Engineering. He has been a university professor since 2004 (full time since 2010) and published more than 75 articles in journals with impact index JCR in the fields of renewable energy (PV, wind, grid-connected or stand-alone systems), electricity storage (advanced battery models) and other. He has more than 40 communications in congresses, most of them international. He is the co-author of four books and seven book chapters, participated in 10 R+D+i projects and 13 contracts and is author of a patent in operation (iHOGA software, web: https://ihoga.unizar.es/en/). He was the director of six doctoral Ph.D. theses, five final master's projects and more than 40 final degree projects. He has spent three months at the Institute of Systems and Robotics (ISR, Portugal, Portugal) and another four months at the Université de Pau et des Pays de l'Addour (UPPA, Pau, France). He has interest and experience in the generation of electricity through renewable sources (off-grid systems and grid-connected systems), advanced optimization techniques and storage of electrical energy (advanced battery models). Interests: renewable energy; electricity storage; advanced batteries models; net metering; energy management; optimization algorithms.

José Luis Bernal-Agustín Electrical Engineer, Ph.D., is University Professor in the Department of Electrical Engineering at the University of Zaragoza. He has published 54 articles in journals with a JCR impact index, having reached an H index of 29. These publications have been referenced on 3325 occasions (without considering self-citations). He was the supervisor of eight doctoral theses, one of which obtained the extraordinary prize of doctorate of the University of Zaragoza, and two mention from the European doctorate. He has also presented more than 70 papers in conferences (most of them international). He is the leading researcher of a European project within the LIFE program, leading researcher of three national projects and Principal Investigator of two research projects financed by the University of Zaragoza-Ibercaja and one by the University of Zaragoza-Banco Santander. He has participated in several Spanish-Portuguese Integrated Actions and several national and regional projects. He is interested in renewable energy, advanced optimization techniques, energy management, electricity markets and optimization techniques.

Editorial

Special Issue on Standalone Renewable Energy System: Modeling and Controlling

Rodolfo Dufo-López * and José L. Bernal-Agustín

Department of Electrical Engineering, Universidad de Zaragoza, C/María de Luna, 3, 50018 Zaragoza, Spain; jlbernal@unizar.es
* Correspondence: rdufo@unizar.es

Received: 12 March 2020; Accepted: 13 March 2020; Published: 19 March 2020

1. Introduction

Standalone (off-grid) renewable energy systems supply electricity in places where there is no access to a standard electrical grid. These systems may include photovoltaic generators, wind turbines, hydro turbines or any other renewable electrical generator. Usually this kind of system includes electricity storage (commonly, lead-acid batteries, but also other types of storage can be used, such as lithium batteries, other battery technologies, supercapacitors and hydrogen). In some cases, a backup generator (usually powered by fossil fuel, diesel or gasoline) is part of the hybrid system.

Low-power standalone systems are usually called off-grid systems and typically power single households by diesel generators or by solar photovoltaic (PV) systems (solar home systems) [1]. Systems of higher power are called micro- or mini-grids, which can supply several households or even a whole village. Mini- or micro-grids, powered by renewable sources, can be classified as smart grids, allowing information exchange between the consumers and the distributed generation [2].

The modelling of the components, the control of the system and the simulation of the performance of the whole system are necessary to evaluate the system technically and economically. The optimization of the sizing and/or the control is also an important task in this kind of systems.

2. Modelling and Controlling Standalone Renewable Energy Systems

Standalone (off-grid) renewable energy systems are used all around the world, and not only in developing countries, as they are the most competitive way to supply electricity in locations where the distance to the transmission and distribution electrical grid is relatively high [3], for example in remote rural communities, farms, telecom stations, etc. Even in some cases, grid-connected systems can become off-grid systems to avoid dependence on the national grid system [4] (however, disconnecting from the grid usually implies higher cost of electricity).

When there is a unique source of energy (for example, solar home systems) the design and optimization of the system is relatively easy. However, the optimal design and operation of the hybrid off-grid systems is a difficult task, as there are many non-linear variables involved which imply that advanced optimization techniques must be used in some cases [5], for example heuristic techniques (genetic algorithms and others). Energy management in mini- and micro-grids with different sources of generation and energy storage is also non-trivial [2,6]. The optimal management of the planning is very important when the system includes fossil-fuel generators (diesel, gasoline) and batteries [7], in order to reduce fuel consumption and enhance battery lifetime.

Usually the main source of energy in the optimal hybrid off-grid system is a photovoltaic generator [8], and also includes in many cases a diesel or gasoline backup generator and battery storage. In windy places, the optimal hybrid off-grid system may also include wind turbines [9].

Especially in cold places, thermoelectric generators that convert thermal energy (for example, waste heat from a stove) into electricity (Seebeck effect) can be part of the optimal hybrid system [10]. However, the use of thermoelectric generators in these kind of applications is still residual.

Nowadays, most off-grid systems installed in the world include storage using lead acid batteries. However, with the recent reduction of the price of lithium batteries, these kind of batteries may be economically feasible in some cases [8,11].

3. Future Standalone Renewable Energy Systems

Although the Special Issue has been closed, more in-depth research of the modelling and controlling of off-grid systems is expected. The use of lithium batteries is expected to be normalized in several years and new battery technologies will emerge. Perhaps thermoelectric generators or other energy sources can be used in off-grid systems in the future.

Acknowledgments: We would like to thank the contributions of the authors' hardworking and professional reviewers. We also thank the editorial team of Applied Sciences, and give special thanks to Stella Zhang, Assistant Editor from MDPI Branch Office, Beijing.

Conflicts of Interest: The authors declare no conflict of interest.

References

1. International Energy Agency. Energy Access Outlook. 2017. Available online: https://webstore.iea.org/download/summary/274?fileName=English-Energy-Access-Outlook-2017-ES.pdf (accessed on 10 February 2020).
2. Vera, Y.E.G.; Dufo-López, R.; Bernal-Agustín, J.L. Energy Management in Microgrids with Renewable Energy Sources: A Literature Review. *Appl. Sci.* **2019**, *9*, 3854. [CrossRef]
3. Mini Grid Policy Toolkit. European Union Energy Initiative Partnership. Dialogue Facility (EUEI PDF): Eschborn, Germany. Available online: http://minigridpolicytoolkit.euei-pdf.org/policy-toolkit.html (accessed on 10 February 2020).
4. Park, E.; Kwon, S.J.; Del Pobil, A.P. Can Large Educational Institutes Become Free from Grid Systems? Determination of Hybrid Renewable Energy Systems in Thailand. *Appl. Sci.* **2019**, *9*, 2319. [CrossRef]
5. Fathima, H.; Palanisamy, K. Optimization in microgrids with hybrid energy systems—A review. *Renew. Sustain. Energy Rev.* **2015**, *45*, 431–446. [CrossRef]
6. Khan, A.A.; Naeem, M.; Iqbal, M.; Qaisar, S.; Anpalagan, A. A compendium of optimization objectives, constraints, tools and algorithms for energy management in microgrids. *Renew. Sustain. Energy Rev.* **2016**, *58*, 1664–1683. [CrossRef]
7. Lujano-Rojas, J.M.; Yusta, J.M.; Artal-Sevil, J.S.; Domínguez-Navarro, J.A. Day-Ahead Optimal Battery Operation in Islanded Hybrid Energy Systems and Its Impact on Greenhouse Gas Emissions. *Appl. Sci.* **2019**, *9*, 5221. [CrossRef]
8. García-Vera, Y.E.; Dufo-López, R.; Bernal-Agustín, J.L. Optimization of Isolated Hybrid Microgrids with Renewable Energy Based on Different Battery Models and Technologies. *Energies* **2020**, *13*, 581. [CrossRef]
9. Dou, Y.; Zuo, G.; Chang, X.; Chen, Y. A Study of a Standalone Renewable Energy System of the Chinese Zhongshan Station in Antarctica. *Appl. Sci.* **2019**, *9*, 1968. [CrossRef]
10. Dufo-López, R.; Champier, D.; Gibout, S.; Lujano-Rojas, J.M.; Domínguez-Navarro, J.A. Optimisation of off-grid hybrid renewable systems with thermoelectric generator. *Energy Convers. Manag.* **2019**, *196*, 1051–1067. [CrossRef]
11. Jung, W.; Jeong, J.; Kim, J.; Chang, D. Optimization of hybrid off-grid system consisting of renewables and Li-ion batteries. *J. Power Sources* **2020**, *451*, 227754. [CrossRef]

© 2020 by the authors. Licensee MDPI, Basel, Switzerland. This article is an open access article distributed under the terms and conditions of the Creative Commons Attribution (CC BY) license (http://creativecommons.org/licenses/by/4.0/).

Article

Day-Ahead Optimal Battery Operation in Islanded Hybrid Energy Systems and Its Impact on Greenhouse Gas Emissions

Juan M. Lujano-Rojas, José M. Yusta, Jesús Sergio Artal-Sevil * and José Antonio Domínguez-Navarro

Department of Electrical Engineering, Universidad de Zaragoza, Calle María de Luna 3, 50018 Zaragoza, Spain; lujano.juan@gmail.com (J.M.L.-R.); jmyusta@unizar.es (J.M.Y.); jadona@unizar.es (J.A.D.-N.)
* Correspondence: jsartal@unizar.es

Received: 24 October 2019; Accepted: 26 November 2019; Published: 30 November 2019

Abstract: This paper proposes a management strategy for the daily operation of an isolated hybrid energy system (HES) using heuristic techniques. Incorporation of heuristic techniques to the optimal scheduling in day-head basis allows us to consider the complex characteristics of a specific battery energy storage system (BESS) and the associated electronic converter efficiency. The proposed approach can determine the discharging time to perform the load peak-shaving in an appropriate manner. A recently proposed version of binary particle swarm optimization (BPSO), which incorporates a time-varying mirrored S-shaped (TVMS) transfer function, is proposed for day-ahead scheduling determination. Day-ahead operation and greenhouse gas (GHG) emissions are studied through different operating conditions. The complexity of the optimization problem depends on the available wind resource and its relationship with load profile. In this regard, TVMS-BPSO has important capabilities for global exploration and local exploitation, which makes it a powerful technique able to provide a high-quality solution comparable to that obtained from a genetic algorithm.

Keywords: vanadium redox flow battery; genetic algorithm; binary particle swarm optimization; time-varying mirrored S-shaped transfer function; greenhouse gas emissions

1. Introduction

Global warming and other environmental problems are driving the adoption of renewable energy sources at the residential, commercial, and industrial levels. Estimating the impact of climate change on the ecosystem involves the accurate knowledge of the carbon cycle and its associated uncertainty. Calculating cumulative emissions in order to prevent an extreme warming level is a key step to guide the manner in which industrial processes, including power generation, should be carried out. Actions for reducing global warming are adjusted following the threshold of 1.5 or 2 °C as the critical limit in a time interval between the years 2000 and 2050 or 2100. However, depending on the established assumptions and scenarios, the risk of experiencing extreme conditions at the middle of the century could be a realistic prospect [1].

Under these circumstances, many countries have been changing their energy mix from a fossil-fuel based one to a renewable-based one, incorporating wind and solar photovoltaic energies, as well as demand response programs [2]. In addition, financial tools such as mutual funds are also implemented to provide economic support for these technologies [3].

As renewable energies are intrinsically variable, the power system requires a high degree of flexibility to effectively manage the uncertainty introduced by these sources, and this could be achieved by implementing demand side management or by installing any type of energy storage system (ESS).

Incorporation of ESS can improve the accommodation of renewable generation while reducing greenhouse gas (GHG) emissions. As an example, incorporation of renewable power combined with ESS in California could reduce carbon dioxide (CO_2) emissions from 90% to 72%, whereas renewable power curtailment reduces from 33% to 9%. In the case of Texas, CO_2 emissions could be reduced from 58% to 54% and renewable power curtailment could be reduced from 3% to 0.3% [4]. Combination of carbon capture and storage devices with conventional generation units is also an option to reduce GHG emissions. However, the combination of renewable generation with ESS can be energetically more effective [5].

Historically, the acquisition costs of a battery energy storage system (BESS) have been considerably high, limiting their economic performance and consequently their mass adoption. However, when very low GHG emissions are required, BESS can be a critical device to achieve such an ambitious goal.

In the case of energy provision for an isolated hybrid energy system (HES), incorporation of BESS becomes profitable due to the fact that the fuel consumption and operating hours of a conventional generator are considerably reduced. In the case of a grid-connected HES, retailing rates and feed-in tariffs as well as favorable resources are crucial for their successful adoption [6].

Heuristic techniques are commonly used to carry out the optimal sizing of a specific HES. Consequently, some of them have been implemented in computational programs such as HOMER Pro® [7], iHOGA® [8], and Hybrid2® [9], among others. Dispatch strategies implemented in most of these tools are based on load following and cycle charging concepts. Load following consists of generating power from conventional units only to satisfy net load (NL), and this approach is frequently suggested in a HES with high share of renewable power, which is much higher than load demand over the year. Conversely, a cycle charging strategy forces conventional generator to operate at its rating power when needed to charge BESS with the remaining energy, so this strategy is frequently implemented when renewable generation is limited [10]. It is important to mention that these strategies do not require any forecast of renewable generation or load demand. However, they are very effective in the management of HES of small scale used on rural electrification projects.

In the case of a HES of larger scale, energy forecasts are frequently employed to optimize the daily operation. This is a topic that has been widely studied and it is the focus of this work. A complete literature review is presented in the next section.

1.1. Literature Review

Management of isolated HES considering the influence of renewable resources and their associated variability has been treated by many authors. In this regard, Li et al. [11] developed a procedure for sizing and management of wind–BESS units. Historical wind power time series is analyzed to estimate the low-frequency component, which is the most prominent one. Using the resulting signal, charging–discharging cycles of BESS are determined considering constant power levels. During the charging period, the power to be provided by the wind–BESS unit is set to the minimum power of low-frequency component within that period. Conversely, power generation is scheduled to the highest power of low-frequency component during discharging periods. In theory, these mechanisms ensure the existence of sequential charging–discharging intervals. However, power dispatch settings could be modified to avoid the charging–discharging cycles at partial level. Other issues related to the wind power forecasting error and BESS lifetime have also been incorporated.

Luo et al. [12] created a model for the operation and sizing of wind–BESS to compensate for the forecasting error. Forecasting error is modeled by using a beta distribution, considering extreme conditions related to pessimistic and optimistic perspectives. BESS dynamic behavior, as well as its lifetime, have been also incorporated.

Mohammadi et al. [13] proposed a day-ahead scheduling model of a microgrid (MG) composed of electrical as well as hydrogen and thermal energy storage technologies. Problem formulation was based on a two-stage stochastic programming approach, while its solution was carried out using an enhanced version of cuckoo optimization algorithm. The high flexibility of the studied configuration results is useful to deal with the fact that thermal and electrical energy consumption are typically not synchronized.

O'Dwyer and Flynn [14] paid special attention to the power system operation on a daily basis, using hourly and sub-hourly time steps, under high renewable energy integration and ESS. According to the reported results, the traditional hourly analysis cannot properly estimate the ramping requirements, the number of starts of conventional generators, as well as the role and potential of ESS on the cycling reduction. Consequently, the interdependence between renewable power curtailment, CO_2 emissions, and the cycling process of thermal units is not accurately described.

Wen et al. [15] presented an enhanced security-constrained unit commitment (SCUC) model, which incorporates BESS to mitigate the negative effects of a sudden contingency and consequently to prevent cascading outages. The methodology was formulated as a two-stage mixed integer programming problem and solved by means of Benders decomposition. The same author in [16] introduced a model based on frequency dynamic constrained unit commitment (UC) able to incorporate wind power uncertainty. Interval optimization approach was combined with mixed integer linear programming (MILP) to determine the appropriate unit schedule.

Nguyen and Crow [17] presented a scheduling model with probabilistic constraints based on stochastic dynamic programming (DP). The proposed BESS-model is inspired by the functioning of conventional fuel-based units. Thus, a detailed cost model was developed considering the electrochemical process of BESS.

Khorramdel et al. [18] proposed a UC model based on cost-benefit analysis, in which a probabilistic analysis based on a here-and-now approach was incorporated. Then, particle swarm optimization (PSO) was implemented in order to minimize total generating costs.

Li et al. [19] developed a framework to quantify the benefits of ESS incorporation to HES. The methodology is based on stochastic UC solved by means of MILP.

Jiang et al. [20] proposed a management model for a residential HES provided by wind generation, micro-combined heat and power generation and smart appliances, enrolled in a real-time pricing (RTP) program. Additionally, optimal behavior of several aggregated HESs is analyzed by means of a day-ahead stochastic economic dispatch (ED) and UC model based on MILP.

Anand and Ramasubbu [21] presented a scheduling model of a system enrolled in a RTP program composed of wind and photovoltaic generation, as well as a microturbine and a fuel cell, based on anti-predatory PSO.

Wu et al. [22] proposed a methodology to solve ED and UC problems using the time-scaling transformation combined with an auxiliary continuous vector.

Dui et al. [23] proposed a two-stage scheduling methodology for BESS performance evaluation. In the first step, UC problem including the effects of thermal and wind generators is solved by means of second-order cone programming. Then, in the second step, the management strategy for BESS is designed and evaluated using a genetic algorithm (GA).

Psarros et al. [24] investigated the operation of HESs using a MILP. BESS sizing is deeply discussed, concluding that this element is a key device for the provision of fast energy reserve. The same author in [25] proposed a model able to consider different time resolutions, based on the combination of model predictive control and MILP.

Ahmadi et al. [26] presented a model for the solution of SCUC including BESS. Aging cost related to BESS operation is incorporated to the objective function. Then, MILP is combined with information-gap decision theory so that the conservatism of the strategy to be implemented can be adjusted by the system operator.

Saleh [27] created and experimentally tested the performance of an energy management system (EMS) based on the solution of UC by Lagrangian relaxation. Thus, control values of permanent magnet generator of the wind turbine and the power-electronic converter are obtained.

Gupta et al. [28] formulated a SCUC model including the effects of BESS in order to compensate the variability of renewable power sources. The mathematical problem is solved by using Benders decomposition, determining the locational marginal price, wind power curtailment, as well as the line contingency.

Alvarez et al. [29] proposed a general purpose ESS model inspired by the behavior of hydraulic reservoirs. Using the results obtained from stochastic dual DP, carried out to determine the long-term energy schedule, the linear model of ESS cost is derived. Finally, stochastic UC, including the aforementioned ESS model, was formulated.

Chen et al. [30] developed a scheduling model based on multi-agent system for the coordination of multiple MGs. Such coordination is carried out by means of the alternating direction method of multipliers, obtaining the optimal energy management of the multiple MGs. Additionally, the negative effects of uncertainty sources are compensated by using day-in rolling.

Tan et al. [31] proposed a dispatch model able to incorporate different operating perspectives related to fuel savings, carbon emissions, power generation costs, amount of renewable energy integrated, and power generation efficiency. Uncertainty of renewable generation and forecasting of carbon-trading price were included by using Monte Carlo simulations, while the associated optimization model was solved by implementing the technique for order preference by similarity to ideal solution combined with Grey relational analysis.

Yiwei et al. [32] presented a scheduling model for a HES based on renewable and thermal generation, as well as cascade hydropower and pumped ESS. The model focuses on the economy and security aspects of system operation. The optimization strategy is divided into three main stages. During the first stage, integer variables are preprocessed using heuristic rules. Then, during the second stage, ED and UC are solved. Finally, during stage three, power system feasibility was evaluated.

Once the literature review has been exposed, describing the state-of-the-art techniques used for day-ahead scheduling, the main contribution and novelty of this work are carefully explained in the next section.

1.2. Main Contributions

As can be observed from the presented literature review, a vast family of methodologies has been created, some of them based on heuristic techniques such as GA and PSO, another group inspired by DP, and most of them based on MILP combined with Benders decomposition.

In a general sense, the optimization technique to be selected strongly depends on the characteristics and assumptions of the ESS model, as well as the context (isolated or grid-connected system) and the information available.

To take advantage of the vast family of BESS models, a recently developed version of binary PSO (BPSO), which incorporates a time-varying mirrored S-shaped (TVMS) transfer function, has been adopted in this paper. Consequently, hourly behavior of charging–discharging efficiency as well as the influence of charge controller on battery operation can be effectively incorporated. Additionally, the influence of wind-speed daily profile on battery schedule and GHG emissions is deeply analyzed. The impact of battery operation on the emissions of total hydrocarbons (THC), carbon monoxide (CO), oxides of nitrogen (NO_X), CO_2, and particulate matter (PM) is investigated.

The remainder of the paper is organized as follows. Section 2 describes the mathematical models of the system configuration under study. Section 3 explains the formulation of the optimization problem and its solution by TVMS-BPSO. Then, problem formulation is tested in Section 4 through a sensitivity analysis based on GA. As TVMS-BPSO is a novel version of BPSO, its performance is compared with GA in Section 5. Finally, conclusions and main findings are discussed in Section 6.

2. Hybrid Energy System Model

The structure of the HES under analysis is shown in Figure 1. On one hand, the diesel generator represents the controllable power source able to provide energy under any circumstance. Thus, energy not supplied (ENS) is neglected. Due to the fact that the diesel generator has important operating costs related to fuel consumption and overhauling, the incorporation of the wind generator combined with BESS and power converter allows us to reduce the number of operating hours of the diesel unit, reducing the operating costs of the whole system.

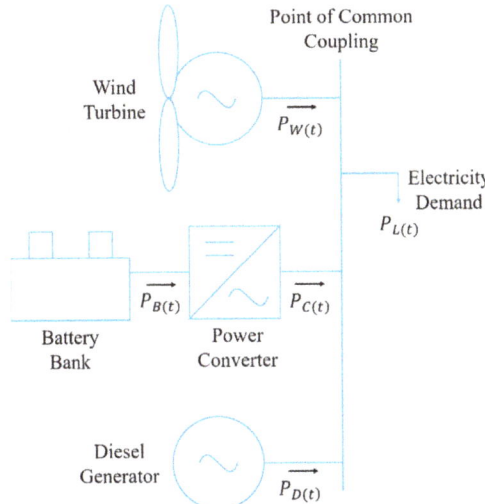

Figure 1. Hybrid energy system (HES) under study.

Besides the wind generator, BESS, power converter, and diesel generator, the dump load (not shown in Figure 1) allows us to consume all the energy surplus of the system in order to maintain the energy balance. This could occur when BESS reaches its maximum capacity and a high magnitude of wind power is available.

In the next sections, computational models of the wind generator, BESS, and diesel generator will be carefully described.

2.1. Wind Generator Model

Wind power generation has been modeled using a typical power curve described according to Equations (1)–(4) [33,34]:

$$P_{W(t)} = \begin{cases} 0; & 0 \leq S_{W(t)} \leq S_W^o \\ P_W^a + P_W^b(S_{W(t)}) + P_W^c(S_{W(t)})^2; & S_W^o \leq S_{W(t)} \leq S_W^r \\ P_W^{max}; & S_W^r \leq S_{W(t)} \leq S_W^f \\ 0; & S_{W(t)} > S_W^f \end{cases} \quad \forall\, t = 1, \ldots, T; \tag{1}$$

$$P_W^a = \frac{1}{\left(S_W^o - S_W^r\right)^2}\left[S_W^o\left(S_W^o + S_W^r\right) - 4 S_W^o S_W^r \left(\frac{S_W^o + S_W^r}{2 S_W^r}\right)^3\right]; \tag{2}$$

$$P_W^b = \frac{1}{\left(S_W^o - S_W^r\right)^2}\left[4\left(S_W^o + S_W^r\right)\left(\frac{S_W^o + S_W^r}{2 S_W^r}\right)^3 - \left(3 S_W^o + S_W^r\right)\right]; \tag{3}$$

$$P_W^c = \frac{1}{\left(S_W^o - S_W^r\right)^2}\left[2 - 4\left(\frac{S_W^o + S_W^r}{2 S_W^r}\right)^3\right]; \tag{4}$$

In this way, the relationship between the wind speed ($S_{W(t)}$) at a determined time step (t) and the corresponding wind power production ($P_{W(t)}$) is clearly established.

2.2. BESS and Power Converter Models

BESS is a crucial device for the appropriate operation of HES because it provides operational flexibility to the whole system. The technology chosen in this work is the vanadium redox flow battery (VRFB) due to its easy scalability, which makes it appropriate for large-scale integration. The mathematical model adopted is shown in Equations (5)–(15), and it has been experimentally tested and validated in [35–37].

Battery voltage ($U_{B(t)}$) and efficiency ($\eta_{B(t)}$) are defined according to charging and discharging processes using Equations (5) and (6), respectively.

$$U_{B(t)} = \begin{cases} U_{B(t)}^{ch}; & P_{B(t)} > 0 \\ U_{B(t)}^{dis}; & P_{B(t)} \leq 0 \end{cases} \quad \forall\, t = 1,\ldots,T; \tag{5}$$

$$\eta_{B(t)} = \begin{cases} \eta_{B(t)}^{ch}; & P_{B(t)} > 0 \\ \eta_{B(t)}^{dis}; & P_{B(t)} < 0 \end{cases} \quad \forall\, t = 1,\ldots,T. \tag{6}$$

During charging process, when battery power ($P_{B(t)}$) is positive, battery voltage ($U_{B(t)}^{ch}$) is related to the state of charge (SOC) ($SOC_{B(t)}$) according to (7), while charging efficiency for voltage ($\eta_{V(t)}^{ch}$) and energy ($\eta_{E(t)}^{ch}$) are related to SOC and battery power as shown in (8) and (9), respectively. Then, global efficiency of charging phenomena ($\eta_{B(t)}^{ch}$) can be estimated using (10).

$$U_{B(t)}^{ch} = \left(U_{ch}^{a} SOC_{B(t)} + U_{ch}^{b} \right) P_{B(t)} + U_{ch}^{c} SOC_{B(t)} + U_{ch}^{d} \quad \forall\, t = 1,\ldots,T; \tag{7}$$

$$\eta_{V(t)}^{ch} = \frac{U_{ch}^{e} T_E \left(SOC_{B(t)} - U_{ch}^{f} \right) + U_{ch}^{g}}{\left(U_{ch}^{h} SOC_{B(t)} + U_{ch}^{j} \right) P_{B(t)} + U_{ch}^{k} SOC_{B(t)} + U_{ch}^{l}} \quad \forall\, t = 1,\ldots,T; \tag{8}$$

$$\eta_{E(t)}^{ch} = \frac{\left(U_{ch}^{m} SOC_{B(t)} + U_{ch}^{n} \right) P_{B(t)} + U_{ch}^{p} SOC_{B(t)} - U_{ch}^{q}}{P_{B(t)}} \quad \forall\, t = 1,\ldots,T; \tag{9}$$

$$\eta_{B(t)}^{ch} = \eta_{V(t)}^{ch} \eta_{E(t)}^{ch} \quad \forall\, t = 1,\ldots,T. \tag{10}$$

During discharging process ($P_{B(t)} < 0$), battery voltage ($U_{B(t)}^{dis}$) and SOC are related according to the linear expression shown in (11). Voltage and energy efficiencies ($\eta_{V(t)}^{dis}$ and $\eta_{E(t)}^{dis}$) depend on battery power and SOC following (12) and (13), respectively. Thus, discharging efficiency ($\eta_{B(t)}^{dis}$) is estimated through the product of these variables ($\eta_{V(t)}^{dis}$ and $\eta_{E(t)}^{dis}$), as suggested in (14).

$$U_{B(t)}^{dis} = U_{dis}^{a} |P_{B(t)}| + U_{dis}^{b} SOC_{B(t)} + U_{dis}^{c} \quad \forall\, t = 1,\ldots,T; \tag{11}$$

$$\eta_{V(t)}^{dis} = \frac{U_{dis}^{d} |P_{B(t)}| + U_{dis}^{e} SOC_{B(t)} + U_{dis}^{f}}{U_{dis}^{g} T_E \left(SOC_{B(t)} - U_{dis}^{h} \right) + U_{dis}^{j}} \quad \forall\, t = 1,\ldots,T; \tag{12}$$

$$\eta_{E(t)}^{dis} = \frac{|P_{B(t)}|}{U_{dis}^{k} |P_{B(t)}| + U_{dis}^{l} SOC_{B(t)} \left(SOC_{B(t)} - 1 \right) + U_{dis}^{m}} \quad \forall\, t = 1,\ldots,T; \tag{13}$$

$$\eta_{B(t)}^{dis} = \eta_{V(t)}^{dis} \eta_{E(t)}^{dis} \quad \forall\, t = 1,\ldots,T. \tag{14}$$

SOC at a determined time interval (t) is defined using (15), which depends on the battery power and efficiency, calculated by following the equations previously described.

$$SOC_{B(t)} = SOC_{B(t-1)} + \int_{t-1}^{t} \left(\frac{P_{B(t)} \eta_{B(t)}}{E_B^{max}}\right) d\tau \ \forall \ t = 1, \ldots, T. \quad (15)$$

Additionally, some operational constrains of VRFB have to be fulfilled. This idea is expressed in (16) for the battery voltage, in (17) for the cell-stack power, and in (18) for SOC:

$$U_B^{min} \leq U_{B(t)} \leq U_B^{max} \ \forall \ t = 1, \ldots, T; \quad (16)$$

$$-P_B^{max} \leq P_{B(t)} \leq P_B^{max} \ \forall \ t = 1, \ldots, T; \quad (17)$$

$$SOC_B^{min} \leq SOC_{B(t)} \leq SOC_B^{max} \ \forall \ t = 1, \ldots, T. \quad (18)$$

Regarding the behavior of power converter, it has been represented through its variable efficiency shown (19), which allows us to estimate the power according to (20).

$$\eta_{C(t)} = \frac{P_{B(t)}}{P_C^a(P_C^{max}) + (1 + P_C^b)P_{B(t)}} \ \forall \ t = 1, \ldots, T; \quad (19)$$

$$P_{C(t)} = \pm \frac{|P_{B(t)}| - P_C^a P_C^{max}}{(1 + P_C^b)} \ \forall \ t = 1, \ldots, T. \quad (20)$$

Regarding the parameters of the VRFB model previously described in (5–15), specifically the parameters $U_{ch}^a - U_{ch}^h$, $U_{ch}^j - U_{ch}^n$, U_{ch}^p, U_{ch}^q for charging and $U_{dis}^a - U_{dis}^h$, $U_{dis}^j - U_{dis}^m$ for discharging; they can be found in [35–37]. Similarly, the parameters P_C^a and P_C^b related to the power converter efficiency have been obtained from the experimental data published in [38].

2.3. Diesel Generator Model

The diesel generator is in charge of satisfying the load that cannot be provided by the wind generator, the battery bank, or both. In addition, this task has to be done considering the technical constraints of the diesel unit. If only the effect of wind generator needs to be considered, NL is calculated according to (21):

$$P_{N(t)} = P_{L(t)} - P_{W(t)} \ \forall \ t = 1, \ldots, T; \quad (21)$$

On the other hand, if the joint effect of the wind generator and BESS needs to be considered, NL can be defined using (22):

$$P_{N(t)} = P_{L(t)} - P_{W(t)} + P_{B(t)} \ \forall \ t = 1, \ldots, T. \quad (22)$$

As aforementioned, the diesel generator has to supply NL as defined in (21) or (22), fulfilling the constraint (23):

$$P_D^{min} \leq P_{D(t)} \leq P_D^{max} \ \forall \ t = 1, \ldots, T. \quad (23)$$

To determine the power dispatch of the diesel unit, the parameter P_D^a is defined according to (24):

$$P_D^a = max(0, P_{N(t)}) \ \forall \ t = 1, \ldots, T. \quad (24)$$

Then, depending on the value of P_D^a, diesel generation ($P_{D(t)}$), power surplus ($P_{EXC(t)}$), and power not supplied ($P_{ENS(t)}$) are determined by following (25–27), respectively,

$$P_{D(t)} = \begin{cases} P_D^{min}; & P_D^a > 0, \; P_D^a \leq P_D^{min} \\ P_D^a; & P_D^a > P_D^{min}, \; P_D^a \leq P_D^{max} \\ P_D^{max}; & P_D^a > P_D^{max} \end{cases} \quad \forall\, t = 1, \ldots, T; \tag{25}$$

$$P_{EXC(t)} = \begin{cases} P_D^{min} - P_D^a; & P_D^a > 0, \; P_D^a \leq P_D^{min} \\ 0; & P_D^a > P_D^{min}, \; P_D^a \leq P_D^{max} \\ 0; & P_D^a > P_D^{max} \end{cases} \quad \forall\, t = 1, \ldots, T; \tag{26}$$

$$P_{ENS(t)} = \begin{cases} 0; & P_D^a > 0, \; P_D^a \leq P_D^{min} \\ 0; & P_D^a > P_D^{min}, \; P_D^a \leq P_D^{max} \\ P_D^a - P_D^{max}; & P_D^a > P_D^{max} \end{cases} \quad \forall\, t = 1, \ldots, T. \tag{27}$$

Once the mathematical model of HES has been defined, the optimization technique proposed in this paper will be clearly explained in the next section.

3. Optimization of Day-Ahead Operation

In this section, the optimization problem and the proposed methodology are carefully described. Section 3.1 pays special attention to the objective function definition, whereas Section 3.2 explains how TVMS transfer function is embedded into BPSO for the daily scheduling of BESS.

3.1. Problem Formulation

The focus of this work is on developing a methodology for load peak-shaving to be applied to the management of autonomous HES. In this regard, EMS monitors the state variables of all the elements connected to the point of common coupling (Figure 1). Then, using this information and the day-ahead forecasts of wind power and load demand, determines how power sources should be dispatched to minimize the operating costs of the system for the corresponding day. Note that the influence of forecasting error on system operation has not been considered in this work.

In a general sense, BESS operation can be defined by means of three different states: charging, discharging, and disconnection. These states can be represented by using integers: charging can be represented as +1, discharging can be represented as −1, whereas 0 represents the battery disconnection.

The goal of the management strategy proposed in this paper consists of finding out the appropriate pattern (charging, discharging, and disconnection) of usage of BESS during the day in order to reduce NL-peak. This is carried out by means of a heuristic optimization algorithm in which each individual or agent is represented as shown in Figure 2. If NL is negative, it means that BESS should be charged in order to store the energy surplus during periods of high wind speed. On the contrary, when NL is positive, it is not evident whether BESS should be discharged or disconnected from HES. Thus, a set of I individuals, who take into account different operational conditions (discharging and disconnection) during different hours, is considered.

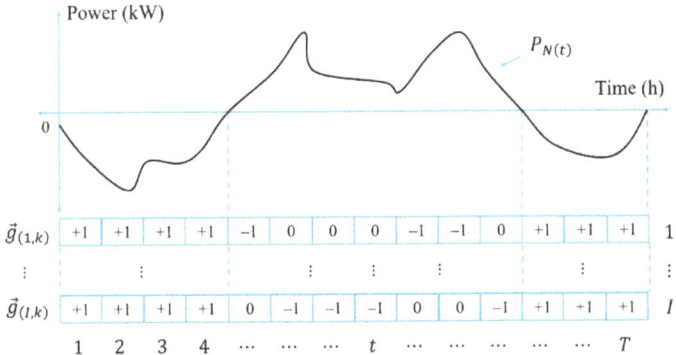

Figure 2. Structure of a single individual.

The structure of a single individual ($i = 1, \ldots, I$) at a determined iteration (k) of the heuristic optimization algorithm can be described according to (28):

$$\vec{g}_{(i,k)} = \begin{bmatrix} g_{(i,1,k)} & \cdots & g_{(i,t,k)} & \cdots & g_{(i,T,k)} \end{bmatrix} \forall\, i = 1, \ldots, I; \tag{28}$$

where each element $g_{(i,t,k)}$ is an integer between -1 and $+1$, depending on the time (t) and NL value ($P_{N(t)}$). Similarly, the population or group of agents of the optimization algorithm for iteration k can be expressed as a matrix according to (29):

$$G_{(k)} = \begin{bmatrix} \vec{g}_{(1,k)} \\ \vdots \\ \vec{g}_{(i,k)} \\ \vdots \\ \vec{g}_{(I,k)} \end{bmatrix}. \tag{29}$$

Considering a determined individual i, the value of its objective function ($O_{(i,k)}$) during a determined iteration (k), which has to be minimized, is calculated according to (30):

$$O_{(i,k)} = \sum_{t=1}^{T} P_{N(t)} P_{C(i,t,k)} \; \forall\, i = 1, \ldots, I; k = 1, \ldots, K; \tag{30}$$

where the pattern $P_{C(i,t,k)}$ corresponds to that obtained from the application of the model previously described in Section 2 (Equation (20)), considering the influence of power converter ($P_{C(i,t,k)} = P_{C(t)}$). In other words, $P_{C(i,t,k)}$ is calculated according to (20), the indexes i and k have been introduced to represent the fact that it is calculated for a specific individual (i) during a determined iteration (k).

The magnitude presented in (30) does not have any physical meaning. Indeed, this has been taken from previous experience of BESS operating in RTP programs [39], where sold and purchased power all over the day is considered as the optimization variable. Following this analogy, $P_{C(i,t,k)}$ can be considered as the transaction power of BESS (sold and purchased power), whereas $P_{N(t)}$ could be considered as linearly related to the fuel-consumption curve of the diesel generator. In other words, in the analogy with selling and purchasing prices under RTP, the variable $P_{N(t)}$ could be considered as a linear function of fuel-consumption costs.

The main conclusion of this reasoning is that, the hour at which BESS should be discharged, in order to maximize profits from energy trading with the RTP scheme, is exactly the same hour at which BESS should be discharged in order to reduce NL-peak in an autonomous HES.

3.2. Optimization by TVMS-BPSO

PSO is an optimization algorithm based on the dynamic behavior of a group of agents interacting with each other. Important variables such as the position ($g_{(i,t,k)}$) and velocity ($v_{(i,t,k)}$) of each agent (i) are considered during the evolution of the algorithm (k). Using constriction factor approach, the agent velocity can be expressed using (31):

$$v_{(i,t,k+1)} = \chi \left[v_{(i,t,k)} + C^a_{PSO} R^a_{PSO} \left(g^{PBEST}_{(t)} - g_{(i,t,k)} \right) + C^b_{PSO} R^b_{PSO} \left(g^{GBEST}_{(t)} - g_{(i,t,k)} \right) \right] \quad (31)$$

where C^a_{PSO} and C^b_{PSO} are coefficients selected so that the condition (32) is fulfilled,

$$\varnothing = C^a_{PSO} + C^b_{PSO}; \; \varnothing > 4. \quad (32)$$

The coefficient $\varnothing$ is then used to calculate the factor χ required in (31),

$$\chi = \frac{2}{\left| 2 - \varnothing - \sqrt{\varnothing^2 - 4\varnothing} \right|}. \quad (33)$$

Using the coefficient $\varnothing$, the convergence of the algorithm can be managed.

TVMS transfer function has been recently proposed by Beheshti [40] to improve the capabilities of BPSO. TVMS-BPSO uses two sigmoid functions during the conversion of reals to binaries, and these functions are shown in (34) and (35):

$$S^a_{PSO(i,t,k+1)} = \frac{1}{1 + e^{\sigma(k)(-v_{(i,t,k+1)})}}; \quad (34)$$

$$S^b_{PSO(i,t,k+1)} = \frac{1}{1 + e^{\sigma(k)(v_{(i,t,k+1)})}}; \quad (35)$$

where the coefficient $\sigma_{(k)}$ varies during the algorithm evolution according to (36):

$$\sigma_{(k)} = (\sigma_{max} - \sigma_{min}) \left(\frac{k}{K} \right) + \sigma_{min}. \quad (36)$$

Once the variables $S^a_{PSO(i,t,k+1)}$ and $S^b_{PSO(i,t,k+1)}$ have been calculated, they are evaluated on (37) and (38) to get the binary variables $J^a_{PSO(i,t,k+1)}$ and $J^b_{PSO(i,t,k+1)}$, which are a preliminary result of the algorithm.

$$J^a_{PSO(i,t,k+1)} = \begin{cases} 1; & R^c_{PSO} < S^a_{PSO(i,t,k+1)} \\ 0; & R^c_{PSO} \geq S^a_{PSO(i,t,k+1)} \end{cases}; \quad (37)$$

$$J^b_{PSO(i,t,k+1)} = \begin{cases} 1; & R^d_{PSO} < S^b_{PSO(i,t,k+1)} \\ 0; & R^d_{PSO} \geq S^b_{PSO(i,t,k+1)} \end{cases}. \quad (38)$$

The definitive result from the conversion of reals to binaries is based on the value of the objective function $O^a_{(i,k)}$ and $O^b_{i,k}$ obtained from the evaluation of $J^a_{PSO(i,t,k+1)}$ and $J^b_{PSO(i,t,k+1)}$ previously estimated. Then, the positions to be considered during the next iteration (k + 1) are defined by using (39):

$$g_{(i,t,k+1)} = \begin{cases} J^a_{PSO(i,t,k+1)}; & O^a_{PSO(i,k+1)} < O^b_{PSO(i,k+1)} \\ J^b_{PSO(i,t,k+1)}; & O^a_{PSO(i,k+1)} > O^b_{PSO(i,k+1)} \end{cases}. \quad (39)$$

Once the principles of TVMS-BPSO have been exposed. The problem of day-ahead BESS scheduling on a daily basis and the TVMS-BPSO performance to solve this problem are analyzed in Sections 4 and 5, respectively.

4. Testing the Problem Formulation

To evaluate the mathematical formulation previously presented in Section 3.1, the performance of HES of Figure 1 is analyzed using a GA and a typical system with a wind generator of 75 kW (P_W^{max} = 75 kW), and the load profile of Figure 3 has been used. Cut-in, rated, and cut-out wind speeds equal to 3, 12, and 25 m/s, respectively, have been assumed.

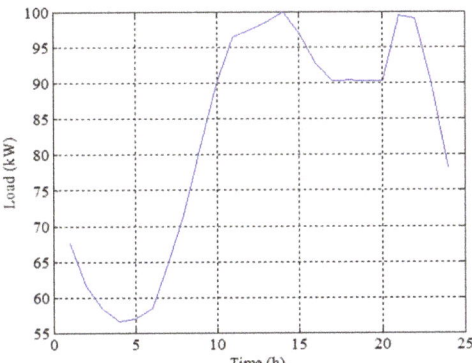

Figure 3. Load profile.

Regarding the fuel-based generator, a diesel unit of 100 kW (P_D^{max} = 100 kW) with minimum operating power of 50% (P_D^{min} = 50 kW) has been assumed.

As the effects of wind generation and BESS management on the reduction of fuel consumption have been deeply studied in the technical literature [41], special attention to the influence of these devices on GHG emissions has been paid in this work. Thus, the GHG-emission measurements published in [42] have been adopted.

During the GA implementation, the initial population is randomly initialized, so that an operator has to be incorporated in order to fix the elements of the matrix $G_{(k)}$ to +1 at those hours at which NL is negative (Figure 2). Additionally, such operator has to be included after the application of mutation operator.

Regarding the GA parameters, a population with 75 individuals, 100 generations, a crossover rate of 90%, and a mutation rate of 5% were considered.

Wind speed profile has been modeled by using the general purposes profile of (40) [7,10], which depends on the average wind speed (S_W^a), diurnal pattern strength (S_W^b), and the hour of peak wind speed (S_W^c).

$$S_{W(t)} = S_W^a \left\{ 1 + S_W^b \cos\left[\left(\frac{2\pi}{T}\right)(t - S_W^c)\right] \right\} \forall\, t = 1, \ldots, T. \tag{40}$$

Different values of average wind speed and diurnal pattern strength have been considered. Specifically, S_W^b = 0, 0.1, 0.2, 0.3, 0.4 and S_W^c = 15 h were evaluated through the analysis of three case studies. These are typical values for places located in the United States. In this way, different values of S_W^b allow us to evaluate the wind speed profile with high or low oscillation, and consequently their impact on the operation of BESS.

A typical VRFB of 5 kW/20 kWh (P_B^{max} = 5 kW/ E_B^{max} = 20 kWh) has been considered. Minimum and maximum SOC were assumed as 15 and 90% (SOC_B^{min} = 0.15 and SOC_B^{max} = 0.9), respectively, and minimum and maximum voltage were assumed as 42 and 56.5 V (U_B^{min} = 42 V and U_B^{max} = 56.5 V), respectively. The entire bank is composed of 10 of these batteries connected in parallel.

The simulation and optimization analysis were implemented in MATLAB®, using a personal computer with i7-3630QM CPU at 2.4 GHz, 8 GB of memory and a 64-bit operating system.

The previously mentioned cases are carefully discussed in the next subsections.

4.1. Case I: Low Wind Speed with Fully Charged Battery

Conditions of low wind speed and fully charged BESS were simulated by considering an average speed of 4 m/s ($S_W^a = 4$ m/s) and an initial SOC equal to 85% ($SOC_{B(0)} = 0.85$).

Figure 4 presents the wind speed (left) and wind power (right) obtained from the evaluation of (40) for the aforementioned values of S_W^b. Because the average speed is close to the cut-in one, high wind speed oscillations ($S_W^b = 0.4$) are directly reflected in the wind power profile.

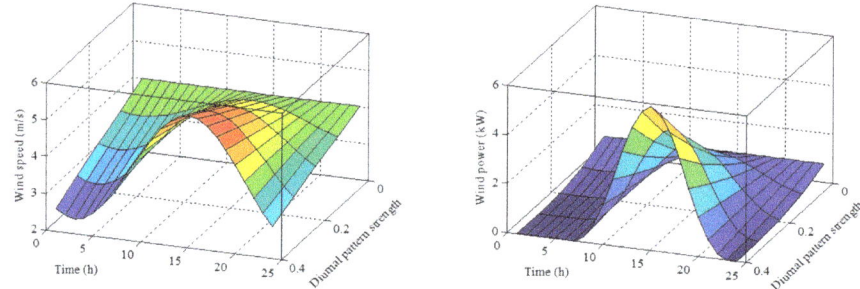

Figure 4. Wind speed and wind power (Case I).

Figure 5 shows the GA-convergence, which takes around 20 iterations to establish a near-optimal schedule.

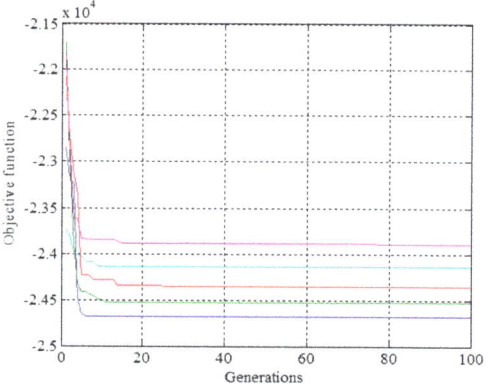

Figure 5. Genetic algorithm (GA) evolution (Case I).

Figure 6 presents the day-ahead schedule of BESS for this case. As can be observed, BESS remains disconnected during the morning, between $t = 1$ h and $t = 10$ h in all cases. Then, BESS is discharged between $t = 11$ h and $t = 15$ h for most of the cases, followed by some disconnection periods, so as to be later discharged during the last hours of the day, between $t = 20$ h and $t = 24$ h. These resting intervals or periods of battery disconnection allow us to improve the management of the stored energy by moving it towards the NL-peak hours.

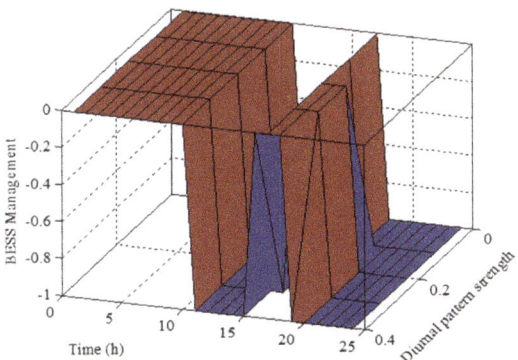

Figure 6. Battery management (Case I).

Figure 7 shows the power (left) and SOC (right) of VRFB, where it is possible to observe how battery power is gradually reduced in order to fulfill the operating conditions (16) and (17) for battery voltage and SOC, respectively.

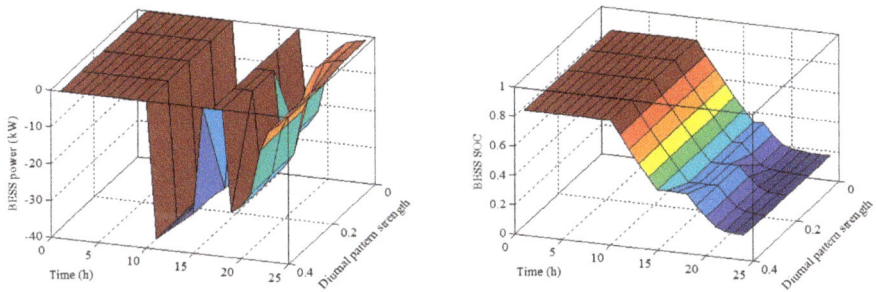

Figure 7. Battery power and state of charge (SOC) (Case I).

Figure 8 presents NL considering the entire architecture of Figure 1 (left) and only considering the wind and diesel generators (right), calculated using (22) and (21), respectively. According to these results, the proposed peak-shaving strategy is effective at discharging the energy initially stored on BESS during those hours of high electricity demand.

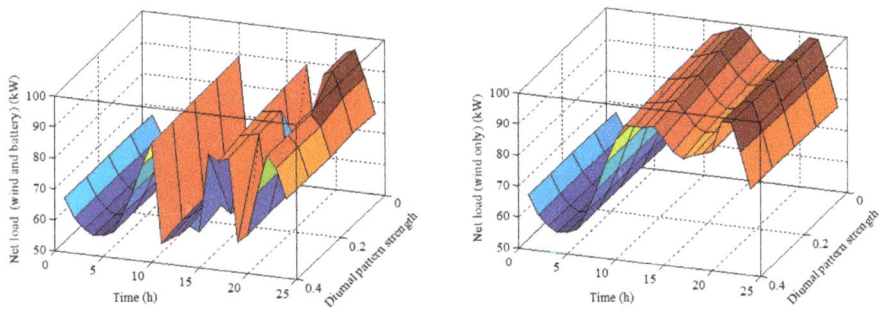

Figure 8. Net load with and without battery (Case I).

Figure 9 presents THC (left) and CO (right) emissions. By comparing NL considering the effect of wind and BESS previously shown in Figure 8 (left) with THC emissions shown in Figure 9 (left), it is possible to observe how NL has a convex shape, whereas THC emissions have concave behavior, which clearly suggests that THC emissions could increase as NL is reduced. Regarding the relationship between NL (Figure 8 left) and CO emissions (Figure 9 right), concave behavior of both surfaces is clearly observed, which means that CO emissions can be reduced with the corresponding limitation of the NL to be supplied by diesel generator.

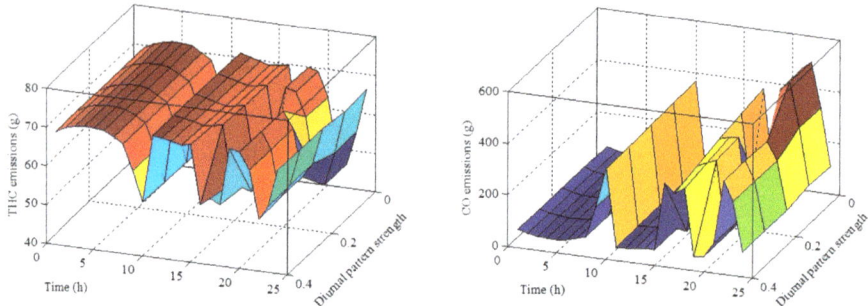

Figure 9. THC and CO emissions (Case I).

Figure 10 shows NO_x (left) and CO_2 (right) emissions, and Figure 11 presents PM emissions. It is possible to observe how all of them slightly increase at the end of the day, due to the fact that BESS management strongly focuses on NL-peak mitigation.

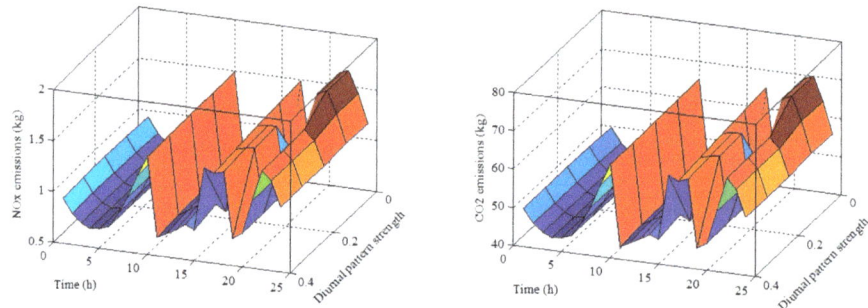

Figure 10. NO_X and CO_2 emissions (Case I).

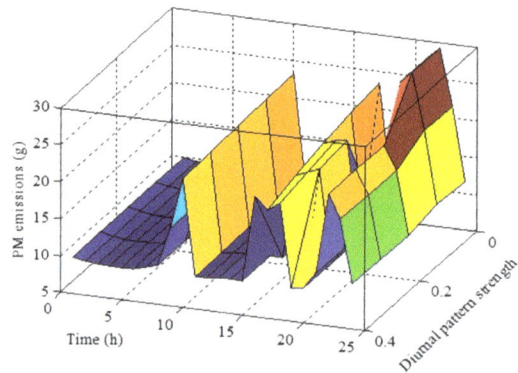

Figure 11. PM emissions (Case I).

Figure 12 verifies that there is no energy surplus (left) or energy not supplied (right) because renewable generation is very low and the diesel generator is able to supply all electricity demand.

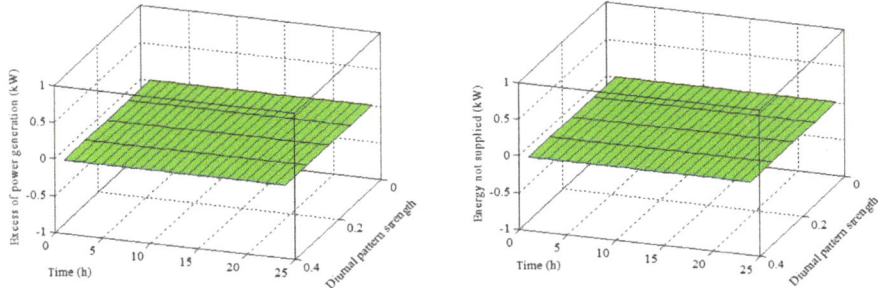

Figure 12. Excess and energy not supplied (ENS) (Case I).

Tables 1–4 summarize the daily results related to GHG emissions and diesel generation. By comparing the second columns of Tables 1 and 2, it is possible to observe how THC emissions increase as previously discussed. In the contrary case, a reduction on CO, NO_x, CO_2, and PM is clearly observed.

Table 1. Total GHG emissions without battery (Case I).

S_W^b	THC (kg)	CO (kg)	NO_X (kg)	CO_2 (kg)	PM (kg)
0	1.37	7.18	31.18	1444.12	0.49
0.1	1.38	7.08	31.13	1442.31	0.48
0.2	1.38	6.94	30.98	1438.01	0.48
0.3	1.39	6.75	30.76	1431.30	0.47
0.4	1.40	6.53	30.47	1422.96	0.46

Table 2. Total GHG emissions with battery (Case I).

S_W^b	THC (g)	CO (g)	NO_X (kg)	CO_2 (kg)	PM (g)
0	1.57	3.51	24.70	1261.25	0.32
0.1	1.57	3.50	24.65	1260.11	0.32
0.2	1.58	3.36	24.57	1257.46	0.32
0.3	1.58	3.30	24.36	1251.68	0.31
0.4	1.58	3.22	24.09	1244.37	0.31

Table 3. Reduction of GHG emissions (Case I).

S_W^b	THC (%)	CO (%)	NO_X (%)	CO_2 (%)	PM (%)
0	−14.43	51.19	20.79	12.66	33.54
0.1	−14.10	50.61	20.79	12.63	32.98
0.2	−13.97	51.54	20.68	12.56	33.37
0.3	−13.48	51.09	20.80	12.55	32.72
0.4	−12.92	50.66	20.95	12.55	32.00

Table 4. Diesel power generation (Case I).

t/S_W^b	Without BESS					With BESS				
	0	0.1	0.2	0.3	0.4	0	0.1	0.2	0.3	0.4
1	66.8	67.3	67.5	67.6	67.6	66.8	67.3	67.5	67.6	67.6
2	60.8	61.3	61.5	61.5	61.5	60.8	61.3	61.5	61.5	61.5
3	57.4	57.9	58.2	58.2	58.2	57.4	57.9	58.2	58.2	58.2
4	55.7	56.2	56.5	56.5	56.5	55.7	56.2	56.5	56.5	56.5
5	56.2	56.6	56.9	56.9	56.9	56.2	56.6	56.9	56.9	56.9
6	57.6	58.0	58.3	58.4	58.4	57.6	58.0	58.3	58.4	58.4
7	63.9	64.2	64.4	64.6	64.6	63.9	64.2	64.4	64.6	64.6
8	71.0	71.2	71.3	71.4	71.5	71.0	71.2	71.3	71.4	71.5
9	80.6	80.6	80.6	80.6	80.6	80.6	80.6	80.6	80.6	80.6
10	89.2	89.0	88.8	88.6	88.3	89.2	89.0	88.8	88.6	88.3
11	**95.7**	**95.3**	**94.9**	**94.3**	**93.7**	58.4	58.0	57.6	57.0	56.4
12	**96.7**	**96.2**	**95.5**	**94.6**	**93.6**	61.6	61.0	60.3	59.5	58.4
13	**97.8**	**97.1**	**96.2**	**95.0**	**93.6**	64.7	64.0	63.0	61.9	60.5
14	**99.2**	**98.4**	**97.4**	**96.0**	**94.4**	68.1	67.3	66.2	64.9	63.2
15	**96.1**	**95.3**	**94.1**	**92.7**	**91.0**	66.9	66.0	64.9	63.4	61.7
16	**91.9**	**91.1**	**90.0**	**88.7**	87.0	64.4	63.6	62.6	61.2	87.0
17	**89.4**	**88.7**	**87.8**	**86.6**	**85.2**	89.4	88.7	87.8	86.6	85.2
18	**89.6**	**89.1**	**88.4**	**87.5**	**86.5**	63.8	63.2	88.4	87.5	86.5
19	**89.4**	**89.0**	**88.6**	**88.1**	**87.4**	67.4	67.0	88.6	88.1	60.0
20	**89.4**	**89.2**	**89.0**	**88.8**	**88.5**	79.5	79.3	63.2	62.9	62.7
21	**98.6**	**98.6**	**98.6**	**98.6**	**98.6**	94.3	94.3	76.6	76.6	76.6
22	**98.2**	**98.4**	**98.5**	**98.6**	**98.7**	95.9	96.0	88.6	88.8	88.9
23	**88.8**	**89.1**	**89.3**	**89.5**	**89.5**	87.2	87.6	85.0	85.2	85.2
24	77.3	77.7	77.9	78.0	78.0	76.1	76.6	75.6	75.7	75.7

Table 3 shows the change on GHG emissions based on the results reported in Tables 1 and 2. The increment of THC emissions is between 12.92% and 14.43%, whereas the reduction of CO, NO$_x$, CO$_2$, and PM are approximately 51.02%, 20.80%, 12.59%, and 32.92%, respectively.

Table 4 shows in bold those hours at which power generation is reduced as a result of BESS incorporation, especially during the peak-load of the afternoon and night.

4.2. Case II: High Wind Speed with Empty Battery

In this case, conditions of high wind speed with an empty battery are studied. Specifically, an average wind speed of 14 m/s (S_W^a = 14 m/s) and an initial SOC of 15% ($SOC_{B(0)}$ = 0.15) are considered. Wind speed and wind power under these conditions are shown in Figure 13, in the left and right sides, respectively, for S_W^b between 0 and 0.4, as previously specified.

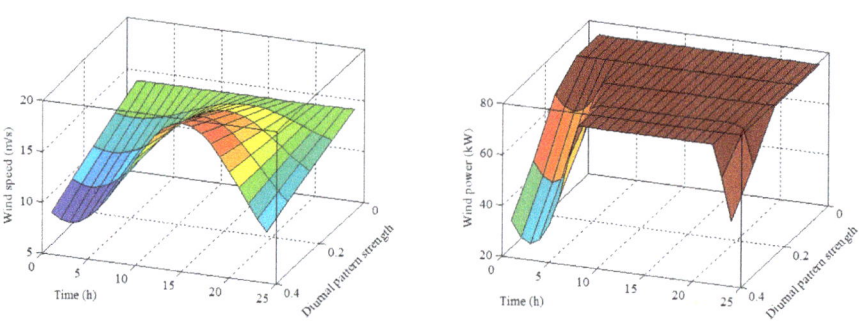

Figure 13. Wind speed and wind power (Case II).

According to the wind power profile (Figure 13 right), rating power is reached in almost all cases, except when a high wind speed oscillation ($S_W^b \to 0.4$) is observed, resulting in a wind power reduction during the morning.

GA evolution is shown in Figure 14, where it can be observed how fast the algorithm converges due to the influence of the high wind speed. In other words, in the presence of high wind speed, the energy surplus forces BESS to be charged, limiting the number of possible operational combinations, and consequently speeding up the convergence.

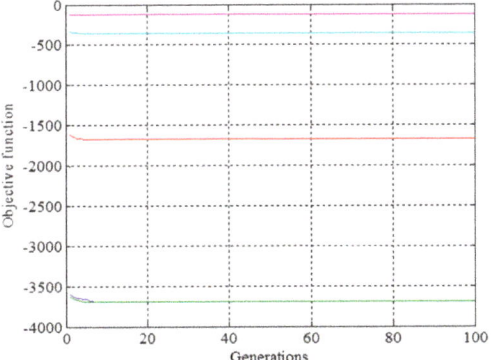

Figure 14. GA evolution (Case II).

BESS management is shown in Figure 15, where the battery is charged during the morning in most situations. However, discharging actions are also advised sometimes in the morning, and this is observed when wind generation reduces as a consequence of wind speed profile oscillations. Initially, the battery is empty, so that discharging the battery in this situation does not result in any load satisfaction, because the battery is not able to provide any power. In other words, battery discharge when $SOC_{B(t)} = SOC_B^{min}$ is equivalent to the battery disconnection obtained by setting $g_{(i,t,k)} \leftarrow 0$.

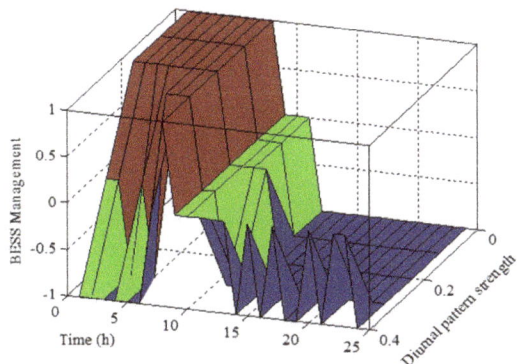

Figure 15. Battery management (Case II).

Figure 16 shows the results obtained from the simulation of the management signal previously shown in Figure 15. Battery power (Figure 16 left) and SOC (Figure 16 right) are presented, and the battery is charged during the morning and then discharged during the afternoon. The situations of high wind speed oscillations result in a reduction of energy surplus, and consequently less power is available to be used in the peak-shaving process during the afternoon.

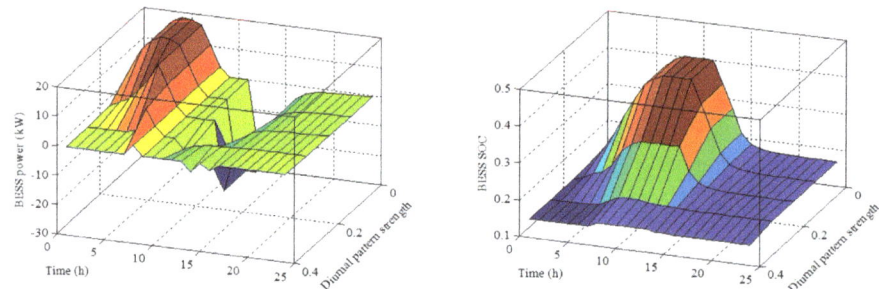

Figure 16. Battery power and SOC (Case II).

Figure 17 shows the NL profile depending on if the effects of BESS are considered (left) or not (right), calculated by using (22) and (21), respectively. By comparing these results, it is possible to observe how NL-peak is reduced.

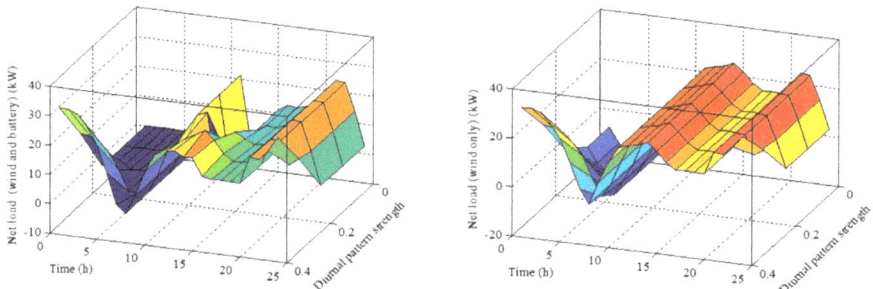

Figure 17. Net load with and without battery (Case II).

As NL is so low when BESS is incorporated, energy surplus is produced because the diesel generator is forced to operate at its minimum capacity in order to satisfy a very low load. This operating mode produces a fixed amount of GHG emissions. This reasoning can be verified in Figure 18 for THC (left) and CO (right), in Figure 19 for NO_X (left) and CO_2 (right), and in Figure 20 for PM.

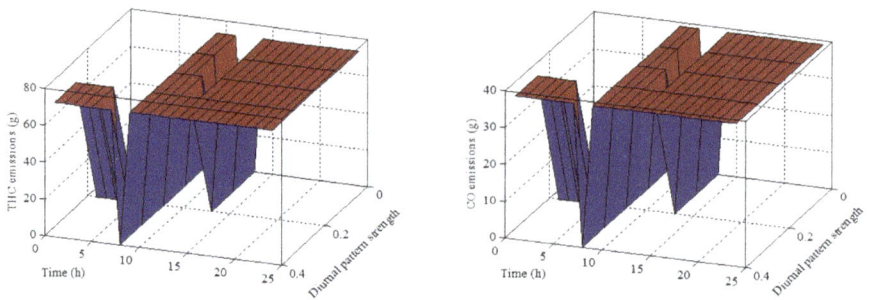

Figure 18. THC and CO emissions (Case II).

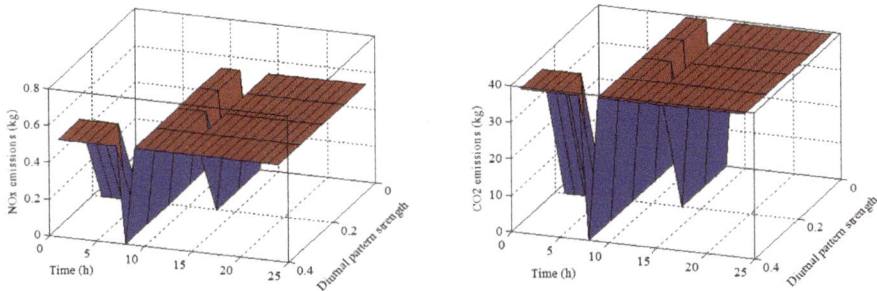

Figure 19. NO$_X$ and CO$_2$ emissions (Case II).

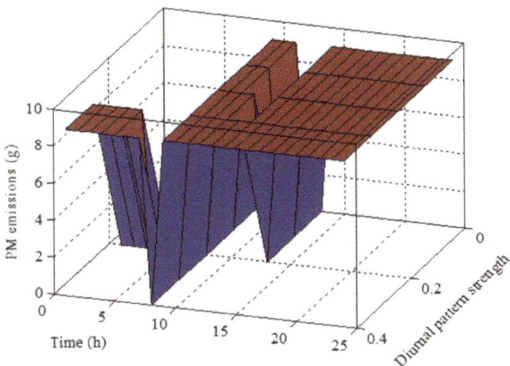

Figure 20. PM emissions (Case II).

Figure 21 shows the daily profiles of energy surplus (left) and ENS (right), respectively. As wind generation is abundant from the high wind speed, energy excess is observed at almost any time. Conversely, there is no ENS.

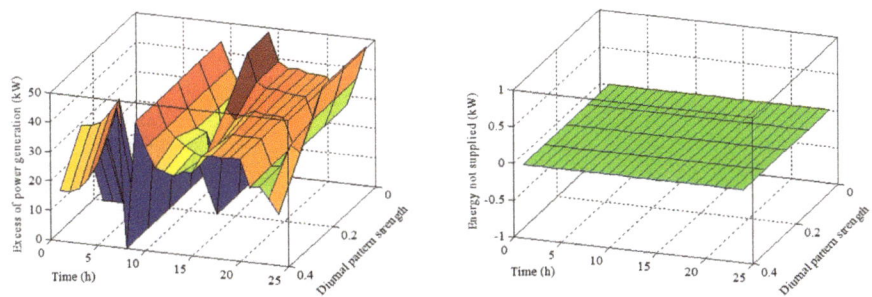

Figure 21. Excess and ENS (Case II).

Tables 5 and 6 report the cumulative daily values of GHG emissions for different wind speed profiles. Then, the reduction on the emitted pollutants as a consequence of BESS integration is reported in Table 7. Because the diesel generator is operating at its minimum allowed power ($P_{D(t)} = P_D^{min}$), the reduction of GHG emissions is the same for all the factors considered, up to 12.5%.

To improve understanding of the HES operation, Table 8 presents the output power of the diesel generator with and without considering the BESS operation. In bold are shown those situations where the diesel unit is disconnected, all of them during the first peak between t = 12 h and t = 13 h.

Table 5. Total GHG emissions without battery (Case II).

S_W^b	THC (kg)	CO (kg)	NO$_X$ (kg)	CO$_2$ (kg)	PM (kg)
0	1.17	0.62	8.48	629.60	0.14
0.1	1.17	0.62	8.48	629.60	0.14
0.2	1.17	0.62	8.48	629.60	0.14
0.3	1.53	0.82	11.13	826.35	0.19
0.4	1.68	0.90	12.19	905.05	0.21

Table 6. Total GHG emissions with battery (Case II).

S_W^b	THC (kg)	CO (kg)	NO$_X$ (kg)	CO$_2$ (kg)	PM (kg)
0	1.02	0.55	7.42	550.90	0.13
0.1	1.02	0.55	7.42	550.90	0.13
0.2	1.10	0.59	7.95	590.25	0.14
0.3	1.53	0.82	11.13	826.35	0.19
0.4	1.68	0.90	12.19	905.05	0.21

Table 7. Reduction of GHG emissions (Case II).

S_W^b	THC (%)	CO (%)	NO$_X$ (%)	CO$_2$ (%)	PM (%)
0	12.5	12.5	12.5	12.5	12.5
0.1	12.5	12.5	12.5	12.5	12.5
0.2	6.25	6.25	6.25	6.25	6.25
0.3	0	0	0	0	0
0.4	0	0	0	0	0

Table 8. Diesel power generation (Case II).

t/S_W^b	Without BESS					With BESS				
	0	0.1	0.2	0.3	0.4	0	0.1	0.2	0.3	0.4
1	0	0	0	50	50	0	0	0	50	50
2	0	0	0	50	50	0	0	0	50	50
3	0	0	0	50	50	0	0	0	50	50
4	0	0	0	50	50	0	0	0	50	50
5	0	0	0	50	50	0	0	0	50	50
6	0	0	0	0	50	0	0	0	0	50
7	0	0	0	0	50	0	0	0	0	50
8	0	0	0	0	0	0	0	0	0	0
9	50	50	50	50	50	50	50	50	50	50
10	50	50	50	50	50	50	50	50	50	50
11	50	50	50	50	50	50	50	50	50	50
12	50	50	50	50	50	0	0	50	50	50
13	50	50	50	50	50	0	0	0	50	50
14	50	50	50	50	50	50	50	50	50	50
15	50	50	50	50	50	50	50	50	50	50
16	50	50	50	50	50	50	50	50	50	50
17	50	50	50	50	50	50	50	50	50	50
18	50	50	50	50	50	50	50	50	50	50
19	50	50	50	50	50	50	50	50	50	50
20	50	50	50	50	50	50	50	50	50	50
21	50	50	50	50	50	50	50	50	50	50
22	50	50	50	50	50	50	50	50	50	50
23	50	50	50	50	50	50	50	50	50	50
24	50	50	50	50	50	50	50	50	50	50

4.3. Case III: Very High Wind Speed with Empty Battery

In this case, conditions of extreme wind speed are analyzed by setting the average speed to 24 m/s (S_W^a = 24 m/s), while the battery remains empty ($SOC_{B(0)}$ = 0.15).

Figure 22 shows the wind speed (left) and wind power (right) for this case. As wind speed becomes higher than the cut-out speed (25 m/s) for those cases with wind speed oscillation, the wind turbine is taken out of service in order to preserve it. Thus, NL suddenly increases during the afternoon.

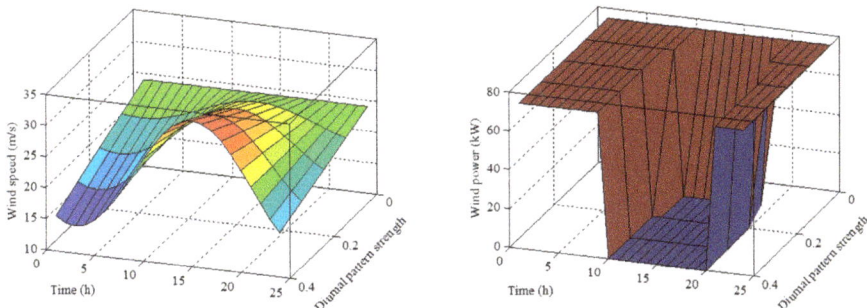

Figure 22. Wind speed and wind power (Case III).

GA evolution is shown in Figure 23, where a fast convergence is observed due to the fact that the battery has to be charged during the first hours of the day, reducing the number of possible combinations of the optimization problem.

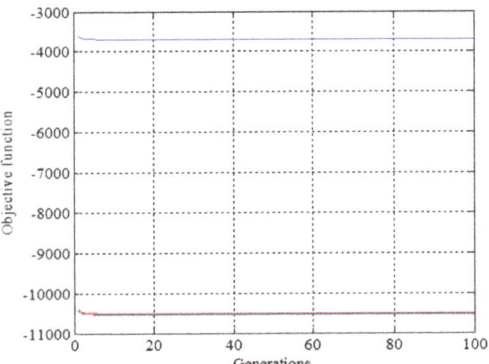

Figure 23. GA evolution (Case III).

Management signal is shown in Figure 24. According to these results, battery should be charged during the morning. Then, a short resting period is advised so that enough energy is available to be discharged during the peak-load hours.

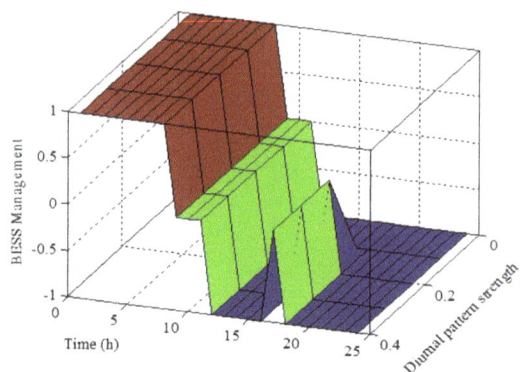

Figure 24. Battery management (Case III).

Charging and discharging cycles are represented by means of the battery power and SOC shown in Figure 25 at the left and right sides, respectively.

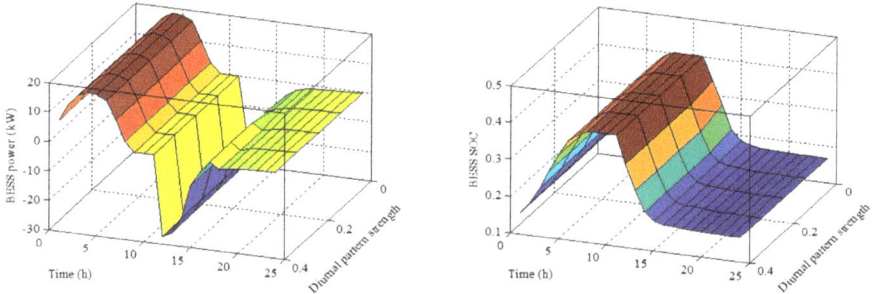

Figure 25. Battery power and SOC (Case III).

The reduction of NL as a result of BESS integration can be observed at the left side of Figure 26. NL without considering the influence of BESS is presented at the right side of Figure 26, where the increment of load demand during the afternoon can be clearly observed.

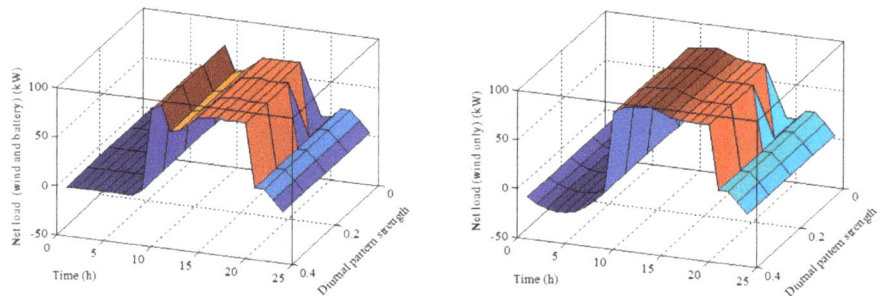

Figure 26. Net load with and without battery (Case III).

GHG emissions are fully described in Figures 27–29, where the lack of wind generation during the afternoon, as a result of the extremely high wind speed, directly influences the behavior of all emission factors.

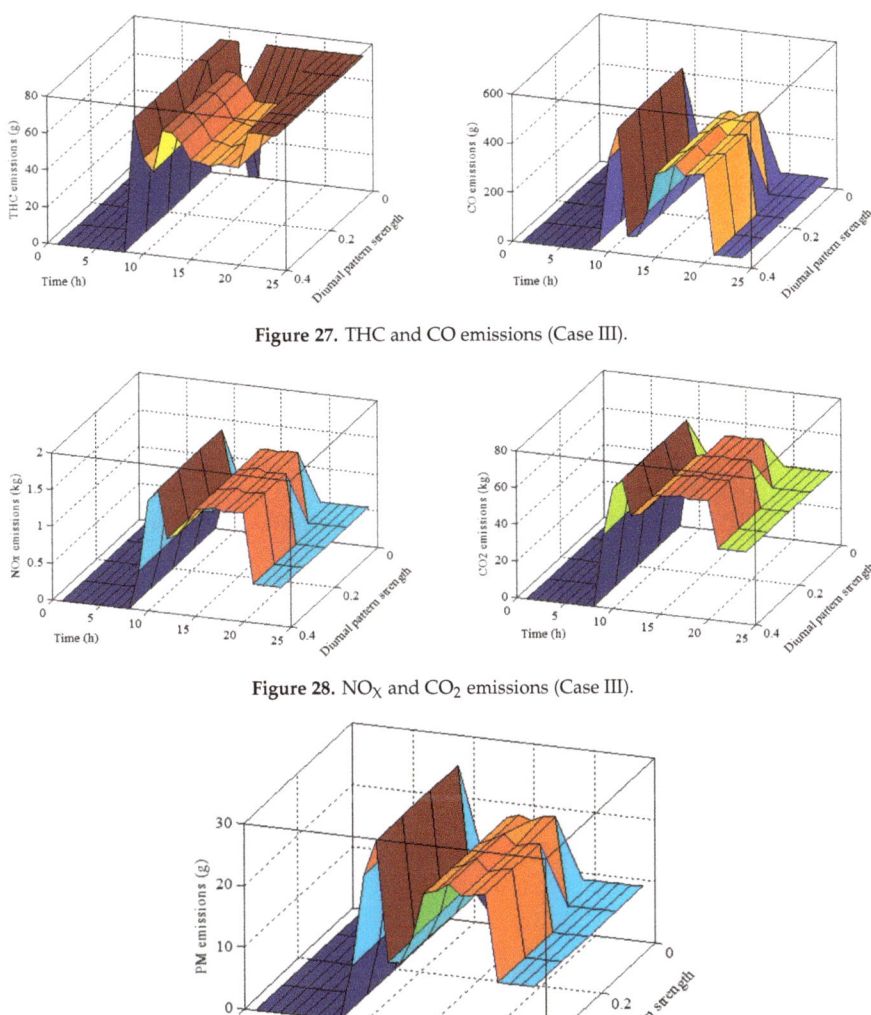

Figure 27. THC and CO emissions (Case III).

Figure 28. NO$_X$ and CO$_2$ emissions (Case III).

Figure 29. PM emissions (Case III).

Energy surplus and ENS are reported in the left and right sides of Figure 30, respectively, where the energy excess is directly related to the operation of the diesel generator during the afternoon. As expected, there is no ENS because the diesel generator is able to supply any value of electricity demand.

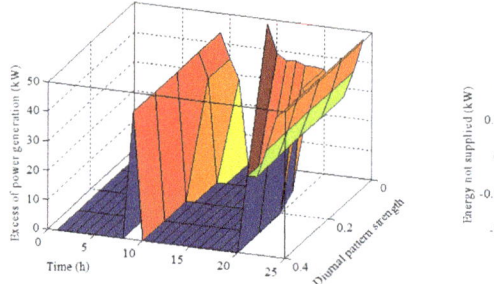 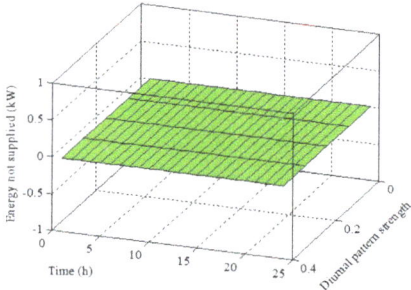

Figure 30. Excess and ENS (Case III).

Tables 9–12 summarize the GHG emissions and diesel generation for this case. By comparing Tables 9 and 10, it is possible to observe that, in some situations, THC emissions can increase when BESS reduces NL in a considerable manner. Regarding the wind speed oscillation, as S_W^b increases, wind speed in the hours close to 15 h ($S_W^c = 15$ h) increases to a value higher than cut-out speed (25 m/s), removing the wind power generation. Thus, BESS supplies a part of the required power, but essentially most of it is provided by the diesel generator, increasing even more the emissions of CO, NO_x, CO_2, and PM to the atmosphere.

Table 9. Total GHG emissions without battery (Case III).

S_W^b	THC (kg)	CO (kg)	NO_X (kg)	CO_2 (kg)	PM (kg)
0	1.17	0.62	8.48	629.60	0.14
0.1	0.94	4.59	18.35	899.93	0.32
0.2	0.90	5.28	20.30	953.02	0.35
0.3	0.90	5.28	20.30	953.02	0.35
0.4	0.90	5.28	20.30	953.02	0.35

Table 10. Total GHG emissions with battery (Case III).

S_W^b	THC (kg)	CO (kg)	NO_X (kg)	CO_2 (kg)	PM (kg)
0	1.02	0.55	7.42	550.90	0.13
0.1	1.02	2.89	16.05	833.17	0.24
0.2	0.98	3.59	18.01	886.27	0.27
0.3	0.98	3.59	18.01	886.27	0.27
0.4	0.98	3.59	18.01	886.27	0.27

Table 11. Reduction of GHG emissions (Case III).

S_W^b	THC (%)	CO (%)	NO_X (%)	CO_2 (%)	PM (%)
0	12.50	12.50	12.50	12.50	12.50
0.1	−9.30	36.98	12.51	7.42	23.93
0.2	−9.73	32.12	11.31	7.00	21.86
0.3	−9.73	32.12	11.31	7.00	21.86
0.4	−9.73	32.12	11.31	7.00	21.86

In Table 12, it can be observed how the diesel generator reduces its power production or is disconnected. This happens in those hours between $t = 12$ h and $t = 20$ h, when BESS has an active role in mitigating NL.

From the sensitivity analysis of Cases I-III, it is possible to conclude that the objective function defined in (30) offers reasonable results in terms of BESS management for peak-shaving. The next section studies the performance of TVMS-BPSO implemented to minimize this already tested objective function.

Table 12. Diesel power generation (Case III).

t/S_W^b	Without BESS					With BESS				
	0	0.1	0.2	0.3	0.4	0	0.1	0.2	0.3	0.4
1	0	0	0	0	0	0	0	0	0	0
2	0	0	0	0	0	0	0	0	0	0
3	0	0	0	0	0	0	0	0	0	0
4	0	0	0	0	0	0	0	0	0	0
5	0	0	0	0	0	0	0	0	0	0
6	0	0	0	0	0	0	0	0	0	0
7	0	0	0	0	0	0	0	0	0	0
8	0	0	0	0	0	0	0	0	0	0
9	50	50	50	50	50	50	50	50	50	50
10	50	50	90	90	90	50	50	90	90	90
11	50	96.4	96.4	96.4	96.4	50	96.4	96.4	96.4	96.4
12	50	97.5	97.5	97.5	97.5	0	70.2	70.2	70.2	70.2
13	50	98.5	98.5	98.5	98.5	0	72.9	72.9	72.9	72.9
14	50	100	100	100	100	50	79.7	79.7	79.7	79.7
15	50	96.9	96.9	96.9	96.9	50	87.9	87.9	87.9	87.9
16	50	92.7	92.7	92.7	92.7	50	88.7	88.7	88.7	88.7
17	50	90.2	90.2	90.2	90.2	50	88	90.2	90.2	90.2
18	50	90.4	90.4	90.4	90.4	50	88.9	88.2	88.2	88.2
19	50	90.2	90.2	90.2	90.2	50	89.1	88.7	88.7	88.7
20	50	50	90.2	90.2	90.2	50	50	89.1	89.1	89.1
21	50	50	50	50	50	50	50	50	50	50
22	50	50	50	50	50	50	50	50	50	50
23	50	50	50	50	50	50	50	50	50	50
24	50	50	50	50	50	50	50	50	50	50

5. Performance of TVMS-BPSO

To evaluate the capabilities of TVMS-BPSO presented in Section 3.2 for day-ahead BESS scheduling, the conditions of Case I (low wind speed with fully charged battery) previously described in Section 4.1 have been considered. This case has been chosen because the number of optimization variables to be determined is the highest. Regarding the number of agents and iterations, these have been set equal to the population size and generations of GA previously implemented in Section 4 (75 agents and 100 iterations), and this guarantees a fair comparison between both methods. Other parameters of BPSO have been adjusted as follows; $C_{PSO}^a = 2.05$, $C_{PSO}^b = 2.05$, $\sigma_{min} = 0.1$, and $\sigma_{max} = 1$.

Figures 31–33 show the comparison between GA and TVMS-BPSO for different wind speed profiles, whereas Table 13 shows the value of the objective function. As can be observed, TVMS-BPSO employs global exploration during the first iterations, analyzing solutions with high objective function value. As the algorithm evolves, exploitation has the relevant role of guiding the algorithm to a high quality solution, comparable to that obtained from GA implementation, according to Table 13.

Table 13. Comparison of objective function values.

S_W^b	GA	BPSO	Difference (%)
0	−24,680.21	−24,679.41	0.00326
0.1	−24,528.93	−24,520.42	0.03469
0.2	−24,348.18	−24,331.04	0.07039
0.3	−24,134.06	−24,116.79	0.07154
0.4	−23,877.54	−23,877.54	0

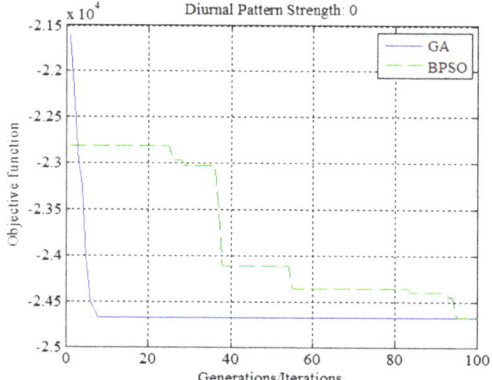

Figure 31. TVMS-BPSO evolution for diurnal pattern strength equal to 0.

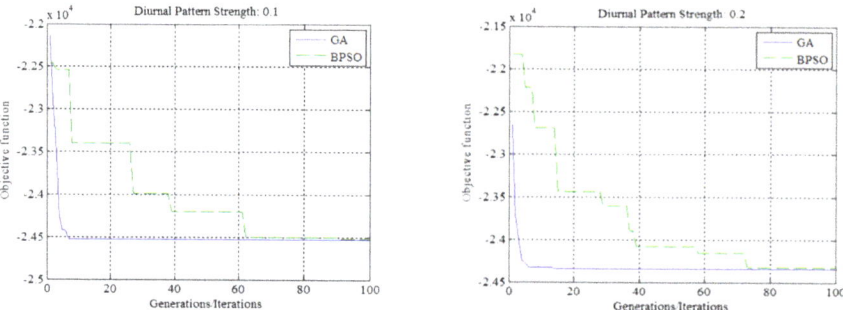

Figure 32. TVMS-BPSO evolution for diurnal pattern strength equal to 0.1 and 0.2.

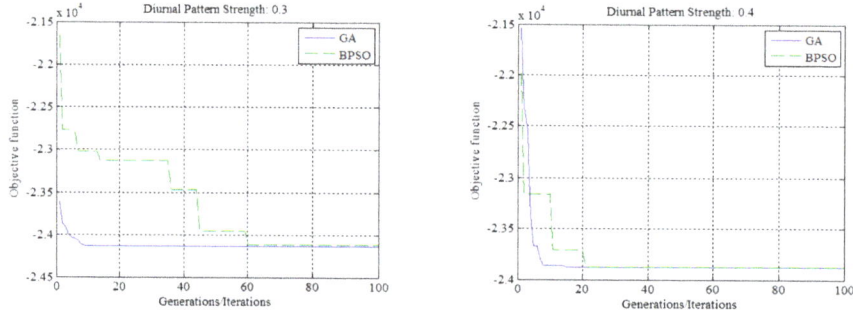

Figure 33. X. TVMS-BPSO evolution for diurnal pattern strength equal to 0.3 and 0.4.

6. Conclusions and Remarks

The results obtained from the analysis of the aforementioned cases offer us important lessons about the mitigation of GHG emissions by integrating a BESS managed from a purely economic perspective. The proposed approach is based on the solution of an optimization problem in which the number of possible combinations varies with the available wind speed profile.

As NL becomes negative, the management signal of BESS is directly set to the charging process ($g_{(i,t,k)} = +1$). In the contrary case, when NL is positive, the optimization approach has to determine whether the BESS should be discharged ($g_{(i,t,k)} = -1$) or disconnected ($g_{(i,t,k)} = 0$) to reduce the daily NL-peak.

Let T be the number of hours during which NL is positive (Case I Section 4.1), the number of decision variables to be determined is defined as 2^T. In Case I, the wind speed profile was so low that no energy surplus was observed. However, the initial SOC was high, so that the proposed approach had to determine how to use that stored energy in order to reduce NL-peak. Under these circumstances, $T = 24$ and the number of possible combinations is 16,777,216. That is why GA and TVMS-BPSO require some iterations to get a near-optimal solution (Figures 5 and 31–33). Conversely, as wind speed increases, as studied in Case II and Case III, energy surplus increases and the number of combinations is reduced, making the management problem easier to solve. This is why an extremely fast convergence is observed in Figures 14 and 23, for Case II and Case III, respectively. With respect to TVMS-BPSO performance, its important capabilities for global exploration and local exploitation offer a high quality solution similar to that obtained from GA implementation (Table 13).

Regarding GHG emissions, the highest reduction was observed in Case I, in which wind power generation was very low, but available energy from BESS was optimally allocated. As Table 3, Table 7, and Table 11 were calculated using the wind–diesel system as reference, Case I presents the highest percentage of reduction. As long as wind speed increases, the diesel generator must be committed to its minimum capacity so that GHG emissions cannot be totally eliminated, and this is important for the optimal sizing of HES. In the presence of an extremely high wind speed, when the wind turbine is disconnected, the reduction of GHG emissions highly depends on how BESS is managed, which can be observed in the results of Case III (Table 11), specifically. Another relevant result is that THC emissions do not always increase with the partial operation of the diesel unit: in Case II, THC emissions were reduced, perhaps because the diesel generator was disconnected in some operational circumstances.

Author Contributions: Conceptualization, J.M.L.-R., J.M.Y., and J.A.D.-N.; Investigation, J.M.L.-R.; Methodology, J.M.Y.; Supervision, J.S.A.-S. and J.A.D.-N.; Validation, J.M.Y. and J.S.A.-S.; Writing—Original draft, J.M.L.-R. and J.S.A.-S.; Writing—Review & editing, J.A.D.-N.

Funding: This work was funded by Ministerio de Economía, Industria y Competitividad of Spanish Government under project number ENE2016-77172-R, by Government of Aragon and the European Union, T28_17R, building Aragon from Europe.

Conflicts of Interest: The authors declare no conflicts of interest.

Abbreviations

i	Index for each individual ($i = 1 : \ldots, I$).
k	Index for each iteration ($k = 1, \ldots, K$).
t	Index for each time step ($t = 1, \ldots, T = 24$).
$S_{W(t)}$	Wind speed at time t (m/s).
S_W^o, S_W^r and S_W^f	Cut-in, rated, and cut-off wind speed, respectively (m/s).
S_W^a	Average wind speed (m/s).
S_W^b	Diurnal pattern strength.
S_W^c	Hour of peak wind speed (h).
$P_{W(t)}$	Wind power at time t (kW).
P_W^{max}	Rated wind turbine power (kW).
P_W^a, P_W^b and P_W^c	Parameters of wind turbine power curve.
T_E	Electrolyte temperature (K).
$U_{B(t)}$	Battery voltage at time t (V).
$\eta_{B(t)}$	Battery efficiency at time t (V).
$SOC_{B(t)}$	Battery state of charge at time t.
$P_{B(t)}$	Battery power at time t (kW).
$P_{C(t)}$	Converter power at time t (kW).
$P_{C(i,t,k)}$	Converter power of individual i at time t and iteration k (kW).
$P_{L(t)}$	Load demand at time t (kW).
$P_{N(t)}$	Net load at time t (kW).

$P_{D(t)}$	Diesel power at time t (kW).
P_D^a	Parameter of diesel power calculation (kW).
$P_{EXC(t)}$	Power surplus at time t (kW).
$P_{ENS(t)}$	Power not supplied at time t (kW).
U_B^{min}, U_B^{max}	Minimum and maximum battery voltage (V), respectively.
SOC_B^{min}, SOC_B^{max}	Minimum and maximum state of charge, respectively.
P_B^{max} and P_C^{max}	Maximum battery and converter power (kW), respectively.
E_B^{max}	Maximum battery capacity (kWh).
P_D^{min}, P_D^{max}	Minimum and maximum diesel power (kW).
$U_{B(t)}^{ch}$	Battery voltage during charge at time t (V).
$U_{ch}^a - U_{ch}^h,$ $U_{ch}^j - U_{ch}^n, U_{ch}^p, U_{ch}^q$	Parameters of battery voltage during charging.
$\eta_{B(t)}^{ch}$	Battery efficiency during charge at time t (V).
$\eta_{V(t)}^{ch}$	Voltage efficiency during charge at time t.
$\eta_{E(t)}^{ch}$	Energy efficiency during charge at time t.
$U_{B(t)}^{dis}$	Battery voltage during discharge at time t (V).
$U_{dis}^a - U_{dis}^h,$ $U_{dis}^j - U_{dis}^m$	Parameters of battery voltage during discharging.
$\eta_{B(t)}^{dis}$	Battery efficiency during discharge at time t (V).
$\eta_{V(t)}^{dis}$	Voltage efficiency during discharge at time t.
$\eta_{E(t)}^{dis}$	Energy efficiency during discharge at time t.
P_C^a, P_C^b	Parameters of power converter.
$\vec{g}_{(i,k)}$	Agent or individual i at iteration k.
$G_{(k)}$	Population or swarm at iteration k.
$O_{(i,k)}$	Objective function of individual i at iteration k.
$\emptyset, \chi, C_{PSO}^a, C_{PSO}^b$	Coefficient of particle swarm optimization.
$R_{PSO}^a - R_{PSO}^d$	Random variables.
$v_{(i,t,k)}$	Velocity of agent i at time t and iteration k.
$g_{(i,t,k)}$	Position of agent i at time t and iteration k.
$g_{(t)}^{PBEST}$	Position of best agent in the group ($i = 1, \ldots, I$) at time t.
$g_{(t)}^{GBEST}$	Position of best agent until the actual iteration (k) at time t.
$\sigma_{(k)}$	Time-varying variable for iteration k.
$\sigma_{min}, \sigma_{max}$	Minimum and maximum value of $\sigma_{(k)}$.
$S_{PSO(i,t,k)}^a, S_{PSO(i,t,k)}^b$	Sigmoid function values for agent i at time t for iteration k.
$J_{PSO(i,t,k)}^a, J_{PSO(i,t,k)}^b$	Binary variables for agent i at time t for iteration k.
$O_{(i,k)}^a, O_{(i,k)}^b$	Objective function values for agent i and iteration k.

References

1. Rogelj, J.; Huppmann, D.; Krey, V.; Riahi, K.; Clarke, L.; Gidden, M.; Nicholls, Z.; Meinshausen, M. A new scenario logic for the Paris Agreement long-term temperature goal. *Nature* **2019**, *573*, 357–363. [CrossRef]
2. Pfeifer, A.; Krajačić, G.; Ljubas, D.; Duić, N. Increasing the integration of solar photovoltaics in energy mix on the road to low emissions energy system—Economic and environmental implications. *Renew. Energy* **2019**, *143*, 1310–1317. [CrossRef]
3. Martí-Ballester, C.-P. Do European renewable energy mutual funds foster the transition to a low-carbon economy? *Renew. Energy* **2019**, *143*, 1299–1309. [CrossRef]
4. Arbabzadeh, M.; Sioshansi, R.; Johnson, J.X.; Keoleian, G.A. The role of energy storage in deep decarbonization of electricity production. *Nat. Commun.* **2019**, *10*, 1–11. [CrossRef]
5. Sgouridis, S.; Carbajales-Dale, M.; Csala, D.; Chiesa, M.; Bardi, U. Comparative net energy analysis of renewable electricity and carbon capture and storage. *Nat. Energy* **2019**, *4*, 456–465. [CrossRef]

6. Comello, S.; Reichelstein, S. The emergence of cost effective battery storage. *Nat. Commun.* **2019**, *10*, 1–9. [CrossRef]
7. HOMER Pro. Available online: https://www.homerenergy.com/ (accessed on 14 October 2019).
8. iHOGA. Available online: https://ihoga.unizar.es/ (accessed on 14 October 2019).
9. Hybrid2. Available online: https://www.umass.edu/ (accessed on 14 October 2019).
10. Lambert, T.; Gilman, P.; Lilienthal, P. Micropower system modeling with HOMER. In *Integration of Alternative Sources of Energy*, 1st ed.; Farret, F.A., Simões, M.G., Eds.; Wiley-IEEE Press: Hoboken, NJ, USA, 2006; pp. 379–418.
11. Li, Q.; Choi, S.S.; Yuan, Y.; Yao, D.L. On the determination of battery energy storage capacity and short-term power dispatch of a wind farm. *IEEE Trans. Sustain. Energy* **2011**, *2*, 148–158. [CrossRef]
12. Luo, F.; Meng, K.; Dong, Z.Y.; Zheng, Y.; Chen, Y.; Wong, K.P. Coordinated operational planning for wind farm with battery energy storage system. *IEEE Trans. Sustain. Energy* **2015**, *6*, 253–262. [CrossRef]
13. Mohammadi, S.; Mozafari, B.; Solymani, S.; Niknam, T. Stochastic scenario-based model and investigating size of energy storages for PEM-fuel cell unit commitment of micro-grid considering profitable strategies. *IET Gener. Transm. Distrib.* **2014**, *8*, 1228–1243. [CrossRef]
14. O'Dwyer, C.; Flynn, D. Using energy storage to manage high net load variability at sub-hourly time-scales. *IEEE Trans. Power Syst.* **2015**, *30*, 2139–2148. [CrossRef]
15. Wen, Y.; Guo, C.; Pandžić, H.; Kirschen, D.S. Enhanced security-constrained unit commitment with emerging utility-scale energy storage. *IEEE Trans. Power Syst.* **2016**, *31*, 652–662. [CrossRef]
16. Wen, Y.; Li, W.; Huang, G.; Liu, X. Frequency dynamics constrained unit commitment with battery energy storage. *IEEE Trans. Power Syst.* **2016**, *31*, 5115–5125. [CrossRef]
17. Nguyen, T.A.; Crow, M.L. Stochastic optimization of renewable-based microgrid operation incorporating battery operating cost. *IEEE Trans. Power Syst.* **2016**, *31*, 2289–2296. [CrossRef]
18. Khorramdel, H.; Aghaei, J.; Khorramdel, B.; Siano, P. Optimal battery sizing in microgrids using probabilistic unit commitment. *IEEE Trans. Ind. Inform.* **2016**, *12*, 834–843. [CrossRef]
19. Li, N.; Uçkun, C.; Constantinescu, E.M.; Birge, J.R.; Hedman, K.W.; Botterud, A. Flexible operation of batteries in power system scheduling with renewable energy. *IEEE Trans. Sustain. Energy* **2016**, *7*, 685–696. [CrossRef]
20. Jiang, Y.; Xu, J.; Sun, Y.; Wei, C.; Wang, J.; Ke, D.; Li, X.; Yang, J.; Peng, X.; Tang, B. Day-ahead stochastic economic dispatch of wind integrated power system considering demand response of residential hybrid energy system. *Appl. Energy* **2017**, *190*, 1126–1137. [CrossRef]
21. Anand, H.; Ramasubbu, R. A real time pricing strategy for remote micro-grid with economic emission dispatch and stochastic renewable energy sources. *Renew. Energy* **2018**, *127*, 779–789. [CrossRef]
22. Wu, X.; Zhang, K.; Cheng, M.; Xin, X. A switched dynamical system approach towards the economic dispatch of renewable hybrid power systems. *Int. J. Electr. Power Energy Syst.* **2018**, *103*, 440–457. [CrossRef]
23. Dui, X.; Zhu, G.; Yao, L. Two-stage optimization of battery energy storage capacity to decrease wind power curtailment in grid-connected wind farms. *IEEE Trans. Power Syst.* **2018**, *33*, 3296–3305. [CrossRef]
24. Psarros, G.N.; Karamanou, E.G.; Papathanassiou, S.A. Feasibility analysis of centralized storage facilities in isolated grids. *IEEE Trans. Sustain. Energy* **2018**, *9*, 1822–1832. [CrossRef]
25. Psarros, G.N.; Papathanassiou, S.A. Internal dispatch for RES-storage hybrid power stations in isolated grids. *Renew. Energy* **2020**, *147*, 2141–2150. [CrossRef]
26. Ahmadi, A.; Nezhad, A.E.; Hredzak, B. Security-constrained unit commitment in presence of lithium-ion battery storage units using information-gap decision theory. *IEEE Trans. Ind. Inform.* **2019**, *15*, 148–157. [CrossRef]
27. Saleh, S.A. Testing a unit commitment based controller for grid-connected PMG-based WECSs with generator-charged battery units. *IEEE Trans. Ind. Appl.* **2019**, *55*, 2185–2197. [CrossRef]
28. Gupta, P.P.; Jain, P.; Sharma, S.; Sharma, K.C.; Bhakar, R. Stochastic scheduling of battery energy storage system for large-scale wind power penetration. *J. Eng.* **2019**, *2019*, 5028–5032.
29. Alvarez, M.; Rönnberg, S.K.; Bermúdez, J.; Zhong, J.; Bollen, M.H.J. A generic storage model based on a future cost piecewise-linear approximation. *IEEE Trans. Smart Grid* **2019**, *10*, 878–888. [CrossRef]
30. Chen, L.; Zhu, X.; Cai, J.; Xu, X.; Liu, H. Multi-time scale coordinated optimal dispatch of microgrid cluster based on MAS. *Electr. Power Syst. Res.* **2019**, *177*, 105976. [CrossRef]

31. Tan, Q.; Ding, Y.; Ye, Q.; Mei, S.; Zhang, Y.; Wei, Y. Optimization and evaluation of a dispatch mode for an integrated wind-photovoltaic-thermal power system based on dynamic carbon emissions trading. *Appl. Energy* **2019**, *253*, 113598. [CrossRef]
32. Yiwei, F.; Zongxiang, L.; Wei, H.; Shuang, W.; Yiting, W.; Ling, D.; Jietan, Z. Research on joint optimal dispatching method for hybrid power system considering system security. *Appl. Energy* **2019**, *238*, 147–163.
33. Giorsetto, P.; Utsurogi, K.F. Development of a new procedure for reliability modeling of wind generating units. *IEEE Trans. Power Appar. Syst.* **1983**, *PAS-102*, 134–143. [CrossRef]
34. Dialynas, E.N.; Machias, A.V. Reliability modelling interactive techniques of power systems including wind generating units. *Archiv Elektrotechnik* **1989**, *72*, 33–41. [CrossRef]
35. Qiu, X.; Nguyen, T.A.; Guggenberger, J.D.; Crow, M.L.; Elmore, A.C. A field validated model of a vanadium redox flow battery for microgrids. *IEEE Trans. Smart Grid* **2014**, *5*, 1592–1601. [CrossRef]
36. Nguyen, T.A.; Qiu, X.; Guggenberger, J.D., II; Crow, M.L.; Elmore, A.C. Performance characterization for photovoltaic-vanadium redox battery microgrid systems. *IEEE Trans. Sustain. Energy* **2014**, *5*, 1379–1388. [CrossRef]
37. Nguyen, T.A.; Crow, M.L.; Elmore, A.C. Optimal sizing of a vanadium redox battery system for microgrid systems. *IEEE Trans. Sustain. Energy* **2015**, *6*, 729–737. [CrossRef]
38. Rampinelli, G.A.; Krenzinger, A.; Romero, F.C. Mathematical models for efficiency of inverters used in grid connected photovoltaic systems. *Renew. Sustain. Energy Rev.* **2014**, *34*, 578–587. [CrossRef]
39. Lujano-Rojas, J.M.; Dufo-López, R.; Bernal-Agustín, J.L.; Catalão, J.P.S. Optimizing daily operation of battery energy storage systems under real-time pricing schemes. *IEEE Trans. Smart Grid* **2017**, *8*, 316–330. [CrossRef]
40. Beheshti, Z. A time-varying mirrored S-shaped transfer function for binary particle swarm optimization. *Inf. Sci.* **2019**. [CrossRef]
41. Lujano-Rojas, J.M.; Dufo-López, R.; Bernal-Agustín, J.L. Technical and economic effects of charge controller operation and coulombic efficiency on stand-alone hybrid power systems. *Energy Convers. Manag.* **2014**, *86*, 709–716. [CrossRef]
42. Shah, S.D.; Cocker, D.R., III; Johnson, K.C.; Lee, J.M.; Soriano, B.L.; Miller, J.W. Emissions of regulated pollutants from in-use diesel back-up generators. *Atmos. Environ.* **2006**, *40*, 4199–4209. [CrossRef]

© 2019 by the authors. Licensee MDPI, Basel, Switzerland. This article is an open access article distributed under the terms and conditions of the Creative Commons Attribution (CC BY) license (http://creativecommons.org/licenses/by/4.0/).

Article

Multi-Objective Optimal Capacity Planning for 100% Renewable Energy-Based Microgrid Incorporating Cost of Demand-Side Flexibility Management

Mark Kipngetich Kiptoo [1],*, Oludamilare Bode Adewuyi [1], Mohammed Elsayed Lotfy [1,2], Tomonobu Senjyu [1], Paras Mandal [3] and Mamdouh Abdel-Akher [4]

1. Graduate School of Science and Engineering, University of the Ryukyus, Okinawa 903-0213, Japan; adewuyiobode@gmail.com (O.B.A.); mohamedabozed@zu.edu.eg (M.E.L.); b985542@tec.u-ryukyu.ac.jp (T.S.)
2. Department of Electrical Power and Machines, Zagazig University, Zagazig 44519, Egypt
3. Department of Electrical and Computer Engineering, University of Texas, El Paso, TX 79968, USA; pmandal@utep.edu
4. Faculty of Engineering, Aswan University, Aswan 81542, Egypt; mabdelakher@ieee.org
* Correspondence: kiptoo.k.mark@gmail.com

Received: 16 August 2019 ; Accepted: 11 September 2019; Published: 13 September 2019

Abstract: The need for energy and environmental sustainability has spurred investments in renewable energy technologies worldwide. However, the flexibility needs of the power system have increased due to the intermittent nature of the energy sources. This paper investigates the prospects of interlinking short-term flexibility value into long-term capacity planning towards achieving a microgrid with a high renewable energy fraction. Demand Response Programs (DRP) based on critical peak and time-ahead dynamic pricing are compared for effective demand-side flexibility management. The system components include PV, wind, and energy storages (ESS), and several optimal component-sizing scenarios are evaluated and compared using two different ESSs without and with the inclusion of DRP. To achieve this, a multi-objective problem which involves the simultaneous minimization of the loss of power supply probability (LPSP) index and total life-cycle costs is solved under each scenario to investigate the most cost-effective microgrid planning approach. The time-ahead resource forecast for DRP was implemented using the scikit-learn package in Python, and the optimization problems are solved using the Multi-Objective Particle Swarm Optimization (MOPSO) algorithm in MATLAB®. From the results, the inclusion of forecast-based DRP and PHES resulted in significant investment cost savings due to reduced system component sizing.

Keywords: demand response program (DRP); photovoltaic system (PV); pumped heat energy storage (PHES); critical peak pricing (CPP) DRP; time-ahead dynamic pricing (TADP) DRP; loss of power supply probability (LPSP); energy storage system (ESS); Multi-Objective Particle Swarm Optimization (MOPSO)

1. Introduction

The quest for provision of affordable, clean, and reliable electricity supply is the key aspiration of many nations globally. These aspirations are portrayed by the commitment of most countries to the formulation and chartering of strategic policies that are targeted towards attaining 100% green energy transition in the near future [1]. Many countries have embarked on different sustainable energy pathways; for example Germany [2] and Sweden [3] aims to attain 100% renewable energy by 2050 while Hawaii in the United States has set 2045 as a target [4]. Several African countries have also taken significant steps and shown visible commitment towards massive green energy uptakes mainly by wind and solar energy. Countries such as Kenya [5], Ghana [6], Mauritius [7], Nigeria [8], Egypt, and

South Africa [9] are currently making efforts in the integration of renewable energy technologies on both small and large scale. However, the incorporation of variable renewable energy resources (VREs) such as wind and solar energy increases the flexibility needs of a power system. Hence, to achieve an acceptable level of power system operation reliability, dynamic and vibrant control strategies need to be devised to balance the demand and supply using efficient flexibility mechanism [10]. Flexibility providers are, usually, adaptable resources needed to address the short-term mismatches between the instantaneous generated power and the load demand [11].

Most of the classical power system planning strategies comprises of segregated optimization models for power system design. These models generally comprise of three features which are component sizing to determine the optimal capacity configuration, unit commitment model to determine the optimal operation strategy, and electricity market strategies to evaluate optimal point-to-point energy transactions [12]. However, the segregated approach is not sufficient for achieving an optimally reliable system design; this is because the operational efficiency of power system relies highly on the time-based dispatchability and controllability of the system generating resources. Furthermore, with the increasing penetration of VREs, the controllability and dispatchability of power system generation sources becomes more complex. Hence, the planning for the transition towards a high VREs-based energy system requires an integrated system planning that involves the cost of component sizing and system flexibility [13]. A comprehensive investigation of the economic viability of different kinds of flexibility providers available for power systems is discussed in [14]. The cost of flexibility is defined as the additional cost required to integrate additional adaptable resources to address the intermittency of VREs integration. There are many sources of flexibility provider options; these includes system interconnection, demand-side management, supply-side management, storage technologies, etc. [15]. From the generation planning perspective, flexibility is investigated based on the ramping capability of the generators, the minimum possible attainable generation, increased cycles of shutdowns and startups for hybrid configuration as outlined in [16].

The idea of hybrid-energy system has also shown some significant growing interest as a valuable and efficient flexibility provider towards 100% VREs generation as shown in much recent research. The authors in [14] performed and provided a comprehensive framework for techno-economic flexibility analysis based on MILP optimization model by combining complimentary distribution generation alternatives such as thermal storage, heat pump, and cogeneration. The importance of the appropriate selection of complementary generating technologies coupled with energy storage system (ESS), with an improved optimal operation strategy, as a cost-effective path towards ensuring power system flexibility is highlighted in [17]. Electricity storage has played a valuable and significant central role in power system in many aspects [18]. Energy storage has the advantage to time-shift the electrical energy supply thus it acts as an ideal mechanism for moderating the consequences of fluctuating output of VREs on the power system. There are many well-known types of ESS in many works of literature varying in terms of technical and economic specification as summarized in [19]. Many studies have evaluated and demonstrated the cost–benefits of appropriate selection and application of different ESS technologies incorporation into the power system planning. The common ESS ranges from pumped hydro [20], hydrogen storage [21], BESS [22], compressed air energy storage, etc. [23]. The inclusion of demand-side management into optimal component sizing that involves energy storage (ESS) facilities is proposed in [24]. The final outcomes show that using the demand-side management (DSM) increases the system flexibility and offers an economical planning option with reduced ESS capacity requirement.

There are two main categories of end-user electrical demand namely the flexible load demand (FDRs) and inflexible/static load demand. The flexible load demand (FDR) are assumed to be those appliances whose time of use can be transferred from one period to another. FDRs include heat pump, room heater, washing machines, etc. They are also referred to as the shiftable appliances because their usage can be delayed during the period of peak demand or shortage of electricity supply and activated later during the period of over-generation. Non-shiftable load demands, on the other hand, are appliances that are static in terms of the time of use, they have a fixed time of period as to when to use, such as illumination loads. DSM has recently received heightened attention in terms of flexibility provision capability using flexible demand resources (FDRs) to achieve the controllability of the customer load demand pattern [25]. In general, the FDRs provide allowances and capacity for time-shifting in terms of their energy requirements. A proper schedule of the FDRs can guarantee mitigation of the gap that exists between demand and system generation for power systems with very high renewable energy fraction as addressed in [26]. A demand-side flexibility approach using the controllability of FDRs has been developed with a detailed implementation framework for commercial and residential smart building in [27]. Demand response program (DRP) is a subset of DSM designed to influences the consumer's behaviors in terms of the time of usage of the FDR through motivations such as incentive payments and lucrative electricity prices to improve the overall system efficiency [28]. The concepts of DRP have been adequately covered in the literature with much focus on optimizing the electricity market design. The commonly featured types of DRPs found in the literature includes and not limited to; real-time pricing, day-ahead pricing, time of use, interrupted curtailable, direct load control, critical peak pricing, etc. [29]. Successful implementation of DRP should take into account the current and the forecasted future power system status to fully exploit the market flexibility [30] and captures the VREs generation uncertainties [31,32].

However, an accurate and reliable VREs output forecasting can serve as a core and vital component of energy management systems (EMS) implementation [33]. The role of forecasting also has significant value in the implementation of pricing schemes in the power markets to decrease the rate of market volatility [34]. Hence, power forecasting plays a pivotal role in flexibility planning for integrating and addressing the uncertainty of the VREs in hybrid power systems. Accurate power forecasting provides critical information of the anticipated status (power shortages and surplus) of the power system ahead of time before the actual occurrence. Hence, a good foresight of the time-ahead generation profile provides an opportunity to plan for future uncertainties adequately and cost-effectively. The ability of a system to meet and handle the growing ramping requirements and volatile residual demand is a significant concern of system operators as the share of wind and solar increases. The economic benefit of accurate solar forecasting in minimizing the generation cost, as well as managing power curtailment was investigated and illustrated by Martinez-Anido et al. [32]. A detailed approach has been adopted for wind power forecast application in [35] and in [36] considering several power market scenarios.

1.1. Research Motivation

Various research has been conducted on optimum component sizing using various optimization techniques to evaluate a cost-effective hybrid microgrid configuration such as PV/biodiesel/BESS using simulated annealing [37], Supercapacitors/BESS/WT/Fuel using Non-dominated Sorted Genetic Algorithm [6], diesel/PV/WT using multi-objective self-adaptive differential evolution algorithm [38], PV/WT/BESS using cuckoo search algorithm [39], MOPSO [40], GA-PSO and MOPSO [41], and more. However, it is observed from the research trends in the literature that in order to ascertain the maximum techno-economic benefits for any microgrid configuration and investment, the flexibility requirements of the system must be factored into its design, i.e., reliability based on adequate system flexibility provision must be prioritized alongside the planning and capacity sizing. Hence, in this study, a multi-objective optimal planning for an isolated microgrid that introduces the cost of flexibility management using ESS and DRP is investigated. The multi-objective design problem is formulated and solved using the Multi-Objective Particle Swarm Optimization (MOPSO) algorithm in MATLAB environment.

1.2. Research Contribution

In view of the above, the main contribution of this work is to introduce a suitable cost-effective framework for incorporating short-term flexibility management requirements into the long-term planning of renewable energy-based microgrid. The total cost of investment and flexibility management, and the supply reliability requirements are investigated and compared under different system design scenarios using the multi-objective optimization approach. The effectiveness of ensembled data-driven renewable energy generation forecasting using the Gradient Boosted regression trees (GBRT) techniques for DSM/DRP flexibility planning and efficient coordination of FDRs is analyzed and compared with the critical peak pricing DRP alternative. The economic advantage of using PHES, compared to BESS, in microgrid applications that requires high renewable energy fraction has been demonstrated through simulation using the data for a real Kenyan microgrid case study.

The rest of the paper is organized as follows; Section 2 presents the methodology and system modeling, Section 3 provides an overview of the FDR, and the techniques of each DRP is described. The optimization problems are formulated in Section 4 while Section 5 provided the details of the case study and simulation parameters, simulation results are outlined and discussed in Section 6 and finally, Section 7 provides the conclusion of the work.

2. Methodology and System Modeling

Figure 1 shows the proposed microgrid system infrastructure; which consist of the WT, PV, PHES, and AC loads connected through an AC bus. The energy management system is also included as the control center for the microgrid. The mathematical models that describe the behavior of each system component and the energy management strategies deployed in this study are discussed below.

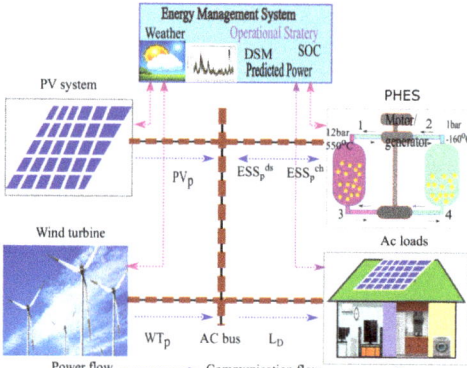

Figure 1. Proposed system model.

2.1. Wind Turbine

The output power of a wind generator $WT_P(t)$ is a function of wind speed and can be calculated using Equation (1) [42]:

$$WT_P(t) = \begin{cases} WT_p^{rtd}(t) \times \frac{u^3 - u_{in}^3}{u_{rtd}^3 - u_{ou}^3} & u_{ci} \leq u \leq u_{rtd} \\ WT_p^{rtd}(t) & u_{rtd} \leq u \leq u_{ou} \\ 0 & u < u_{in} \text{ or } u > u_{ou} \end{cases} \quad (1)$$

where u_{ci}, u_{rtd}, u and u_{ou} are the cut-in speed, nominal speed, instantaneous wind speed at hub height and cut-out wind speed for the wind turbine, respectively. WT_p^{rtd} is the rated power output of the wind turbine.

2.2. PV System

The generated output power of the PV system ($PV_p(t)$) is significantly determined by the solar irradiances incident on the PV surface and temperature. The PV output power as a function of input variables is given by (2) [43]:

$$PV_p(t) = f_{pv} \times \frac{G(t)}{G_{std}} \times [1 + \theta_i(t_{pv}(t) - t_{std})] \times PV_p^{rtd} \qquad (2)$$

where f_{pv}, PV_p^{rtd}, $G(t)$ is the power reduction factor, installed capacity of the PV in kW and the incident solar irradiance, respectively. θ_t, G_{std}, t_{std} is the temperature coefficient, solar irradiance, and temperature under the standard test condition.

2.3. Energy Storage System Model

Whenever the combined output power of WT and PV generation surpasses the capacity of load demand, the ESS transitions into the charging state. The amount of energy stored at any given time t is primarily determined by the difference between the sum of the total PV and WT generation, and the load demand.

2.3.1. Battery Energy Storage System (BESS)

The amount of discharging and charging power drawn or sent to the battery energy storage system, respectively, is subject to the previous state of charge (SOC) as well as the ESS system constraints. The SOC at any given t is determined by the following equation.

$$SOC(t) = \left[(PV_p(t) + WT_p(t)) - \frac{L_D(t)}{\beta_c}\right] \times \beta_{ch} + SOC(t-1)(1-dr) \qquad (3)$$

where $SOC(t-1)$ and $SOC(t)$ and is the BESS state of charge for the previous and current period in kWh, respectively. $L_D(t)$ is the load demand, β_c denotes the power converters efficiency, dr and β_{ch} is the hourly self-discharge rate and BESS charging efficiency respectively. Whenever the total generation cannot meet the load demand, BESS shifts into the discharging mode. Consequently, the current state of charge at any given time t is given by:

$$SOC(t) = \left(\frac{L_D(t)}{\beta_{ds}} - (PV_p(t) + WT_p(t))\right)/\beta_{ds} + (SOC(t-1)(1-dr)) \qquad (4)$$

where β_{ds} is the discharging efficiency. The energy storage level (SOC) must be constrained within the upper SOC_{max} and the lower SOC_{min} bounds of the BESS.

$$SOC_{min} \leq SOC(t) \leq SOC_{max} \qquad (5)$$

2.3.2. Pumped Heat Energy Storage (PHES)

The PHES stores electricity as sensible heat in two thermal storage system; a hot high pressure and temperature tank (+500 °C, 12 bars pressure) and a cold low pressure and temperature tank (−160 °C, 1 Bar). It also consists of a two compressor/expander pair, argon as a working fluid and it uses gravel as the storage medium. The operation strategy is analogous to pumped hydro storage but rather than pumping water, heat pumping is used to create temperature difference. Theory of operation and development is adequately covered in [44–46]. Figure 2. shows the schematic diagram of the PHES.

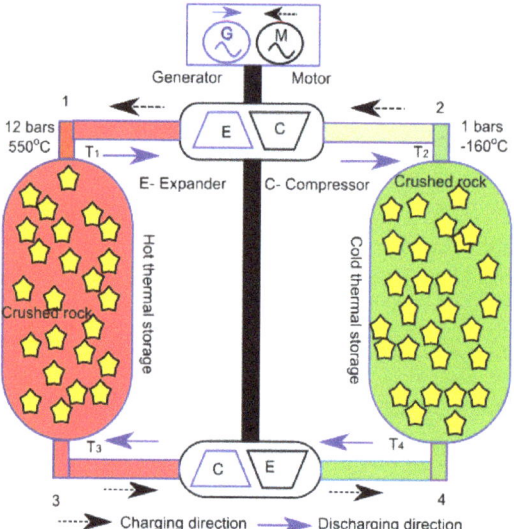

Figure 2. Schematic diagram of PHES.

The energy stored in a PHES depends on the temperature differences between the two thermal storage system. The energy stored $ESS_{phes}(t)$ in the reservoirs per unit volume is the difference between the internal energies of the storage medium in the hot and cold stores. The internal energies of the storage medium are the function of the mass (M_r) and specific heat densities of the storage medium (SH_r). The energy stored can be determined by the temperature difference between the hot and cold store [47] as illustrated below:

$$ESS_{phes}(t) = M_r \times SH_r \times \{(T_2(t) - T_3(t)) - (T_1(t) - T_4(t))\} \quad (6)$$

The power output and input $P_{phes}(t)$ of the PHES per unit volume (for charging and discharging instance) is determined by the mass (Mg) and the specific heat of the argon gas (SHg), and the temperature difference [48] as follows:

$$P_{phes}(t) = M_g \times SH_g \times \{(T_2(t) - T_1(t)) - (T_3(t) - T_4(t))\} \quad (7)$$

where $(T_1(t), T_2(t))$ are the top and $(T_3(t), T_4(t))$ are the bottom section temperature of the hot tank and cold tank respectively.

3. Flexible Demand Resources (FDRs) and Demands Response Program (DRPs)

Figure 3 shows the flowchart for the integrated system planning method considered in this work. The framework combines the optimal ESS scheduling and optimal DRP implementations. The FDRs play significant roles in the flexibility management of the system whenever they are appropriately activated to minimize the mismatch between generation and demand. The DSM approach that is employed in this study for the DRPs is based on the optimal scheduling of appropriate FDRs in the microgrid as explained below. The net capacity of the shiftable load demand (FDR), throughout the system scheduling period, is assumed to have a maximum range of up to 10% up (FDR^{max}) and down (FDR^{min}) of the initial total FDR load demand value [49].

$$FDR^{min} \leq FDR(t) \leq FDR^{max} \quad (8)$$

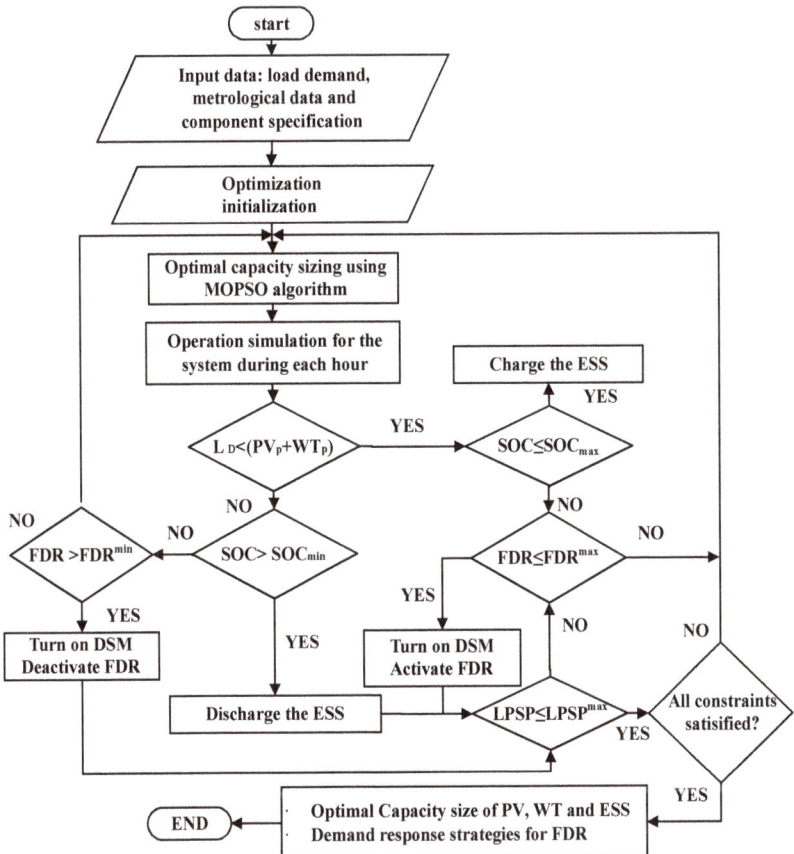

Figure 3. Flowchart for the proposed integrated system planning framework.

3.1. Price Elasticity of Demand and Load Modeling

A change in price of a service will have an impact on the amount of quantity demanded. For instance, a change in the price of electricity ($\partial E_{std}^{pr}(i)$) in the ith period will result in a change of the load demand ($\partial L_D(j)$) in the jth period either by increasing or decreasing the load demand. Thus, a change in electricity price during the single period ith affects the load demand during all the periods (T). The price elasticity of demand ($PE_{\phi(i,i)}$) gives a measure of the responsiveness at which the end-user time-shift their energy consumption patterns with respect to change in electricity as shown below:

$$PE_{\phi(i,i)} = \frac{E^{pr}(i)}{L_D(i)} \cdot \frac{\partial L_D(i)}{\partial E^{pr}(i)}; \quad \forall i,j \in T \tag{9}$$

The price elasticity of demand entails self and cross-elasticity; the self-elasticity defines the sensitivity of load demand with respect to price within the same pricing interval (single period elasticity) and usually has a negative value implying some proportion of the load cannot be transferred from one period to another. On the other hand, cross-elasticity ($PE_{\phi(i,j)}$) defines the load demand sensitivity of the (ith) pricing period in response to the electricity price variation in the (jth) pricing period (multi-period elasticity) and usually has positive value implying some proportion of the load demand is shiftable to another period. The cross-elasticity of load demand is given by [50];

$$PE_{\phi(i,j)} = \frac{E_{std}^{pr}(i)}{L_D(j)} \cdot \frac{\partial L_D(j)}{\partial E^{pr}(i)}; \quad \forall i,j \in T \qquad (10)$$

3.2. Critical Peak Pricing (CPP) Demand Response Program

CPP is a time-based DRP that divide electricity usage time into periods and presents the fixed electricity prices for each period in advance; peak and off-peak periods. It is usually employed to increase system energy efficiency and alleviate stress on the power system especially when the load demand is likely to surpass the generation capacity. It commonly enforces a very high electricity price during system peak load demand periods and for some specific time periods in order to achieve load reduction during these periods, and retains a flat pricing scheme or a lower electricity price during off-peak periods [51]. The electricity customer responds by shifting load demand from one time period to another due to the enforced pricing scheme. The ultimate customer's demand profile after implementation of CPP DRP is expressed as [41,52]:

$$L_D^{cpp}(i) = L_D(i) \left\{ 1 + PE_{\phi(i,i)} \frac{[E_{cpp}^{pr}(i) - E_{std}^{pr}(i) + pd(i) + ps(i)]}{E_{std}^{pr}(i)} \right. \\ \left. + \sum_{j=1, j \neq i}^{T} PE_{\phi(i,j)} \frac{[E_{cpp}^{pr}(j) - E_{std}^{pr}(j) + pd(j) + ps(j)]}{E_{std}^{pr}(j)} \right\}; \quad \text{for all } i,j \in T \qquad (11)$$

where $E_{std}^{pr}(i)$ is the standard Kenyan electricity price before CPP DRP implementation, $E_{cpp}^{pr}(i)$, $E_{cpp}^{pr}(j)$ is the electricity price for current ith period and the jth period after implementation of CPP DRP, $pd(i)$ and $pd(j)$ are the incentives and $ps(i)$ and $ps(j)$ are penalties enforced for non-compliance's of DRP.

3.3. Time-Ahead Dynamic Pricing (TADP) Demand Response Program

The cost of generation and the corresponding cost of electricity are highly affected by the shortages and surplus of power generated in the power system. Short periods of mismatch in load demand and generation might necessitate an over-sizing or additional capacity in the ESS that might not be necessary or efficiently used during normal operating times. A remedy to this challenge is to offer motivating electricity prices to influence a time shift in FDRs by the end user. A longer pricing horizon ahead of time can guarantee end-user participation in the DRP. Thus, in TADP DRP, time-ahead electricity pricing profile formulated as a function of the mismatch in the forecasted demand and generated power is relayed to the end user an hour (one period) in advance.

3.3.1. Time-Ahead Dynamic Pricing Model

The electricity price for the next hour ($E_{TADP}^{pr}(t+1)$) is determined based on the difference between forecasted total generation output power from renewable energy sources (PV and WT) and the load demand $\hat{L}_D(t+1)$ using the following equation:

$$E_{TADP}^{pr}(t+1) = E_{std}^{pr}(t+1) \left(1 + \frac{\hat{L}_D(t+1) - (\hat{WT}_p(t+1) + \hat{PV}_p(t+1))}{\hat{L}_D(t+1)} \right) \qquad (12)$$

where $\hat{PV}_p(t+1)$ and $\hat{WT}_p(t+1)$ represent the forecasted generation output power from the PV and WT, respectively. $E_{std}^{pr}(t+1)$ is the initial (standard) Kenyan electricity price initially present for hour $t+1$ before TADP DRP implementation. $E_{TADP}^{pr}(t+1)$ is the next hour electricity price after the implementation of TADP DRP.

3.3.2. Time-Ahead Dynamic Pricing Demand Response Program Load Modeling

The electricity price determined one hour ahead of time for a specific period is the actual price that would be adopted for that period. Based on the electricity price relayed in advance, the end-consumers are either motivated/discouraged to shift their FDRs. The final economic load model after the TADP DRP is implemented is expressed as:

$$L_D^{TADP}(i) = L_D(i) \left\{ 1 + PE_{\phi(i,i)} \frac{[E_{TADP}^{pr}(i) - E_{std}^{pr}(i) + pd(i) + ps(i)]}{E_{std}^{pr}(i)} + \sum_{j=1, j \neq i}^{T} PE_{\phi(i,j)} \frac{[E_{TADP}^{pr}(j) - E_{std}^{pr}(j) + pd(j) + ps(j)]}{E_{std}^{pr}(j)} \right\}; \quad \text{for all } i,j \in T \quad (13)$$

where $E_{std}^{pr}(i)$ is the current (ith period) Kenyan electricity price before TADP DRP implementation.

3.3.3. Gradient Boosted Regression Trees (GBRT) Model for Time-Ahead Forecast of Generation

In this work, the forecasting tasks are treated as regression problems and machine learning regression algorithms on scikit-learn package in Python are adopted to build the models using the Gradient boosted regression trees (GBRT) algorithm. The GBRT algorithm has a superior advantage of not requiring complex data pre-processing of dimension transformations or reduction and does not suffer any loss of input variable interpretation [53]. The significant feature of the accurate implementation of the GBRT algorithm is the parameter α_{gbr} called the learning rates. The learning rate is a scaling parameter that determines the individual contribution of each decision tree to the final ensemble model. The accuracy of the model is continuously improved by fitting the residual decision iteratively until the desired model is obtained for the best learning rate. Algorithm 1 illustrates the GBRT pseudo code algorithm.

Algorithm 1: Gradient boosted regression trees (GBRT) pseudo code algorithm.

Start:
1. **Precondition**: Input the training data set $M = (m_i, o_i); i = 1..n$ and a differentiate loss function $L_f(o_i, \delta)$
2. **Initialization**: Initialize the model with a constant value:
   $$F_0(m) = argmin \sum_{N}^{i=1} L_f(o_i, \delta)$$
3. **Estimation:** for $i = 1...k$; grow k trees
   (i) Calculate the Pseudo residuals;
   $$r_{ik} = - \left[\frac{\partial L_f(o_i, \delta)}{\partial F(m_i)} \right]_{F(m)=F(m_i)} \quad i = 1..n$$
   (ii) Fit a residual value regression decision tree $I(m)$
   and establish the terminal leaves
   for $J = 1...j_K$; determine the output of each leaves that minimizes;
   $$\delta_{jk} = argmin \sum_{m_i \in R_{ij}} L_f(o_i, F_{k-1}(m_i) + \delta)$$
4. **Update:**
   $$F_k(m) = F_{k-1}(m) + \alpha_{gbr} \sum_{j=1}^{jk} \delta_{jk} I(m \in R_{jm})$$
   End: For
5. **Output** $F_k(m)$

End: Terminate the Algorithm

In order ascertain the accuracy of forecasting algorithms, three performance evaluation metrics are used: Mean Absolute Error (MAE), Root mean squared error (RMSE) and Coefficient of Determination (r^2).

4. Optimal Design Problem Formulation

The multi-objective optimal design model is evaluated in terms of economic and reliability criteria as presented in the objective functions defined below:

4.1. Economic Criteria: Total Life-Cycle Cost (TPC)

The objective function of the economic criterion is formulated as a cost minimization problem of the net present value of the total life-cycle costs (TPC) of all system components alongside the implementation of the flexibility requirements under different system scenarios. The decision variable of the optimization problem is the capacity of the WT (C_{WT}), PV (C_{PV}) and ESS (C_{ESS}).

$$\text{minimize } TPC = \sum_{z=1}^{Z} \left\{ CI_z + \sum_{n=1}^{n=N} \frac{(O\&M_z + RP_z - RV_z)}{(1+r)^n} \right\} \times C_z \quad (14)$$

where z indexes the z_{th} component and C_z is the decision variables that represent the optimum component capacities of each of the system components (PV, ESS, and WT). The TPC components are the capital costs (CI_z), yearly operation and maintenance costs ($O\&M_z$), replacement costs (RP_z) and the salvage value RV_z, N is the project lifetime, n is the time step in the project life, i.e., a year and r is the discount rate. The system components have a yearly operation and maintenances cost over the project lifetime.

4.2. Reliability Criteria: Loss of Power Supply Probability (LPSP)

The second objective considers the loss of power supply probability as the system reliability criteria. $LPSP$ reliability index measures and ascertains the quality and reliability performance of the power system design under the different scenarios considered in this study. $LPSP$ is defined as the ratio of the sum of all energy deficits (LPS) to the total power demand. Thus, $LPSP$ can be evaluated by using the following expression:

$$LPSP = \frac{\sum_{t=1}^{T} LPS(t)}{\sum_{t=1}^{T} L_D(t)} \quad (15)$$

where

$$LPS(t) = L_D(t) - \left[WT_p(t) + PV_p(t) + (SOC(t-1) - SOC_{min}) \times \beta_c \right] \quad (16)$$

$LPSP$ value ranges between zero and one; a value of 0 for $LPSP$ implies that the load demand will always be met or satisfied, and this is the most desired and preferred performance. The following system DRP constraints are considered during the optimization procedure, alongside the other system component constraints that are mentioned at each design stage.

$$\begin{aligned} PV_p(t) + WT_p(t) + ESS_p^{ds}(t) - ESS_p^{ch}(t) &= L_D(t); & \text{without DRP} \\ PV_p(t) + WT_p(t) + ESS_p^{ds}(t) - ESS_p^{ch}(t) &= L_D^{CPP}(t); & \text{with CPP DRP} \\ PV_p(t) + WT_p(t) + ESS_p^{ds}(t) - ESS_p^{ch}(t) &= L_D^{TADP}(t); & \text{with TADP DRP} \end{aligned} \quad (17)$$

4.3. Overview of the Optimization Tool: Multi-Objective Particle Swarm Optimization

PSO is a population-based approach for solving discrete and continuous optimizations problem that stemmed from and mimic the navigation behavior of swarms of bees, flocks of birds, and schools of fish. To obtain the optimal value of the objective function at each search, two different solution points are obtained which are called the local best, $Pbest_i = (p_{i1}, p_{i2}, ..., p_{id})$ and the global best, is $Pbest_g = gbest = (p_{g1}, p_{g2}, ..., p_{gd})$; and the positions of the particles for the next objective function evaluation is estimated as given below:

$$V_{id}^{t+1} = w \times v_{id}^k + c_1 \times rand_1 \times (Pbest_{id} - X_{id}) \\ + c_2 \times rand_2 \times (gbest_d - X_{id}) \tag{18}$$

$$X_{id}^{k+1} = X_{id}^k + V_{id}^{k+1} \tag{19}$$

$$w = w_{damp} \times \frac{iter_{max} - iter}{iter_{max}} + w_i \tag{20}$$

$iter$ is the iteration count, $iter_{max}$ is the total iterations. w_i, w_f are the minimum and maximum range of the inertia weight. The multi-objective PSO approach adopted in this work is described [54]. The repository particles guides the search within the *efficient, non-inferior* and *admissible* pareto front by sorting out the non-dominated solutions. The exploratory capacity of the algorithm is strengthened by a special mutation operator just like in NSGA II algorithm as explained below. If $\vec{f}(\vec{x})$ consists of n objective functions each with m decision variables, then the multi-objective problem can be defined as finding the vector $\vec{x}^* = [x_1^*, x_2^*, ..., x_m^*]^T$ which minimizes $\vec{f}(\vec{x})$ as shown:

$$\text{minimize } \vec{f}(\vec{x}) = [f_1(\vec{x}), f_2(x), ...f_n(\vec{x})] \text{ for } \vec{x}^* \in \varepsilon \tag{21}$$

$$\vec{g}(\vec{x}) \leq 0 \tag{22}$$

$$\vec{h}(\vec{x}) = 0 \tag{23}$$

$\vec{g}$ and $\vec{h}$ are sets of inequality and equality constraints, respectively. A point $\vec{x}^* \in \chi$ is pareto optimal if for every $\vec{x} \in \chi$ and $I = 1, 2, ..., k$ either:

$$\forall i \in I(f_i(\vec{x}) = f_i(\vec{x}^*)) \tag{24}$$

or at least there is one $i \in I$ such that

$$f_i(\vec{x}) > f_i(\vec{x}^*)) \tag{25}$$

5. Research Case Study and Simulation Parameters

The proposed energy system planning and management approach are investigated on an undeserved Marsabit county isolated microgrid in Kenya, which is currently served by conventional diesel-based generators. The goal of this work is to investigate the best flexibility management incorporated hybrid VRE energy supply combination that will completely replace the existing diesel generators considering the cost and reliability criteria that are described above. The hourly meteorological data of the locality (2.3369° N, 37.9904° E) was obtained from online sources [55,56] for 2015 to 2018. The meteorological data set consists of wind speed, wind direction, air pressure, relative humidity, solar irradiance, and the temperature variables. The economic and technical parameters were obtained from [57] through desk research and consultation with energy sector employees and policymakers in the region. Table 1 shows the details of simulation parameters and Table 2 shows the considered self and cross-price elasticity of demand, which is adopted from [52] modified to fit the Kenyan case. The price elasticity of demand entails self and cross-elasticity; the self-elasticity defines the sensitivity of demand with respect to price within the same pricing interval while cross-elasticity ($PE_{\phi(i,j)}$) define the load demand sensitivity of the (ith) pricing period in response to the electricity price variation in the (jth) pricing period. The cross-elasticity of demand is given by [50];

Table 1. Technical, Cost & lifetime parameters of the system components.

System Component and Economic Indicators Specifications		
Economics		
Discount rate	4%	
Inflation rate	3%	
Lifetime of the project	20	years
Specification of the PV system		
Capital costs	1691.5	US $/kW
O & M costs	26	US $/kW/yr
PV reduction factor	0.85	
lifetime	20	years
Specification of the PHES		
Round trip efficiency	70%	
Power converters (Expander/compressor system)	400	US $/kW
Energy storage unit	15.08	US $/kWh
O&M (Power converters units)	12.76	US $/kW/yr
O&M (Thermal Energy unit)	0.03	US $/kWh/yr
self-discharge rate (hourly) d_r	0.04%	
Lifetime	20	years
Specification of the WT		
Capital cost	2030	US $/kW
O& M	76	US $/kW/yr
lifetime	20	years
Wind speed (Cut-in):	4	m/s
Rated wind speed:	14.5	m/s
Cut-out wind speed:	25	m/s
Wind Shear Coefficient	0.143	
Hub height:	50	m
Specification of the BESS		
Capital Cost	300	US $/kWh
O & M	10	US $/kWh/yr
Round trip efficiency	85%	
lifetime	5	years

Table 2. DRP self and cross-price elasticities of demand [49,50,52].

	Off-Peak Period	Peak Period
Off-peak period	−0.1	0.016
Peak period	0.016	−0.1

The Kenyan tariff structure of 2018 was obtained from [58,59]. The current electricity rate of 15.80 US Cents per kWh for ordinary domestic consumers was considered to be the flat rate E_{std}^{pr}. For this work, the CPP DRP pricing scheme was considered to be 20.00 US Cents per kWh for peak period from 7:00 p.m. to 10:00 p.m. while the rest of the day adopted a flat pricing of 15.80 US Cents per kWh. TADP DRP implemented a time-ahead hourly variable pricing scheme with a maximum and minimum electricity price of 20.00 US Cents per kWh and 10.00 US cents per kWh, respectively.

PHES has no geographical limitations [60] and have been found to be a viable ESS technology option for both large and small-scale energy management applications. Its prospects in terms of cost-effectiveness and flexibility provision has also been verified in [48], thus, it has been determined to be one of the most suitable ESS options for application in isolated places such as the Kenyan microgrid case under study. PHES stores electricity as sensible heat in thermally insulated and closed-looped thermal storage systems which ensures that the system is isolated; hence, based on the design aspects outlined in [47], there is a guarantee that the model is feasible for deployment for our case study. The analysis of a proposed commercial PHES design with a maximum capacity of 16 MWh as detailed in [45,61] has been adopted as the benchmark for many studies in the literature; thus, the system technical and economic specifications are used in our work for the Kenyan microgrid under study.

6. Simulation Results and Discussion

The simulation results are presented for three cases based on optimal capacity planning and flexible operation feasibility using BESS and PHES, with and without DR. The optimal size of system components is determined under each case at minimum investment costs and maximum supply reliability (minimum LPSP) while satisfying the system operational and flexibility requirements. The results of the considered three case simulation scenarios are outlined and discussed below:

- Case 1: Comparing BESS and PHES without DRP consideration.
- Case 2: Comparing BESS and PHES with CPP DRP consideration.
- Case 3: Comparing BESS and PHES with TADP DRP consideration.

6.1. Case 1: BESS versus PHES without DRP

Figure 4a,b shows the trade-off Pareto front plots for economic and reliability criteria with BESS and PHES, respectively, under case 1. From Figure 4, as expected, the system reliability condition improves (LPSP value decreases) as the total cost increases and vice visa. Hence, the cost-benefit relationship at different LPSP values is analyzed and discussed using the investment cost-savings approach. Table 3 summarizes the details of the cost-benefit analysis for case 1. The optimal selected points are derived after multiples execution of the optimization program for LPSP values in the range of 0% to 15%.

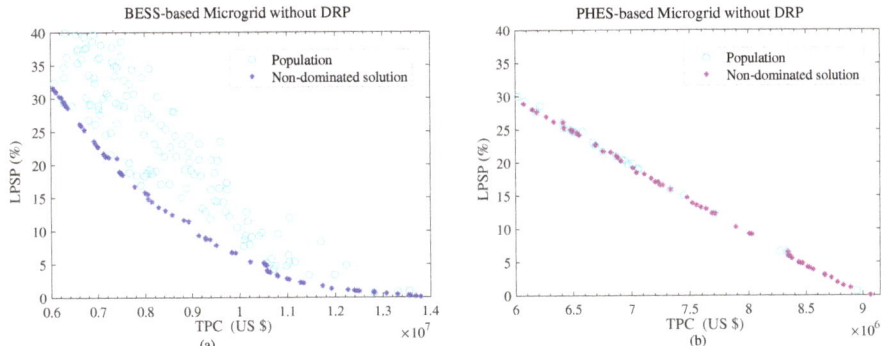

Figure 4. Pareto front plots for case 1.

Table 3. Techno-economic analysis for case 1.

	BESS-Based Microgrid				PHES-Based Microgrid			
LPSP	0%	5%	10%	15%	0%	5%	10%	15%
PV capacity (kW)	1422	1207	1003	1030	1699	1806	1652	1850
WT capacity (kW	1919	2120	2063	1774	1657	1376	1371	1108
ESS capacity (kWh)	6798	1800	900	411	7800	7546	6967	6925
TPC (US $)	1.38×10^7	1.05×10^7	9.28×10^6	8.06×10^6	9.06×10^6	8.38×10^6	8.03×10^6	7.59×10^6

A comparison of the two systems based on the ESS technology at maximum reliability condition i.e., LPSP = 0%; it can be seen that the choice of PHES instead of BESS results in a total investment cost reduction of about 34.28% from US $ 1.38×10^7 to the US $ 9.06×10^6. This a significant cost saving in the microgrid planning. Hence, PHES has been shown more economical compared to BESS.

6.2. Case 2: BESS versus PHES with CPP DRP

In this case, the benefit of CPP DRP on capacity sizing optimization problem for both BESS and PHES-based microgrid is investigated, and pareto fronts plotted. Figure 5a,b shows the Pareto front plots with CPP DRP considering BESS and PHES, respectively. For both cases, it can be observed from the pareto plots that an increase in the LPSP value, the TPC decreases, this is due to the fact that the reliability index (LPSP) and planning cost (TPC) are conflicting objective. Table 4 summarizes the cost–benefits analysis for case 2 which involves the economic effects of critical peak pricing (CPP) DRP for the BESS and PHES-based microgrid configuration.

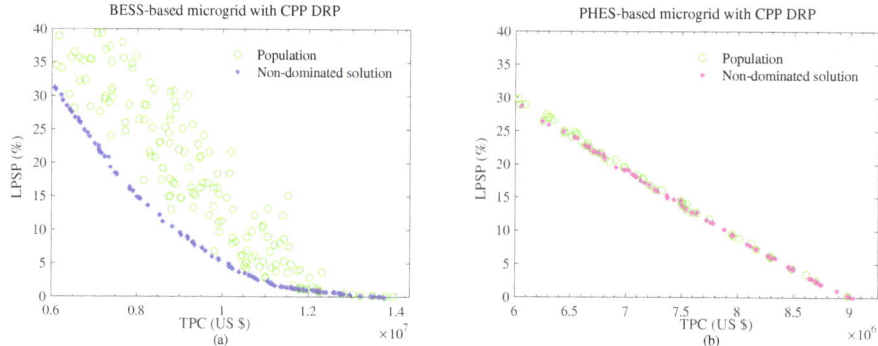

Figure 5. Pareto front for case 2.

Table 4. Techno-economic analysis for case 2.

	BESS-Based Microgrid				PHES-Based Microgrid			
LPSP	0%	5%	10%	15%	0%	5%	10%	15%
PV capacity (kW)	1110	1027	958	1039	1826	1670	1754	1671
WT capacity (kW)	2165	2184	2054	1808	1561	1498	1230	1195
ESS capacity (kWh)	6436	1170	670	132	7789	7311	6986	6414
TPC (US $)	1.37×10^7	9.91×10^6	9.00×10^6	7.99×10^6	9.02×10^6	8.48×10^6	7.78×10^6	7.49×10^6

For the comparative analysis of the two systems configurations at LPSP = 0% (maximum reliability) with the consideration CPP DRP; the selection of PHES as an ESS alternative to BESS in optimum capacity resulted in 34.22% reduction in the total investment costs. This significant cost saving signifies that PHES-based configuration is more economical and preferred investment option compared to BESS-based microgrid.

6.3. Case 3: BESS versus PHES with TADP DRP

In this case, the prospects of TADP DRP in optimum component-sizing problem has been investigated. The renewable energy generation forecasting is a subset feature of the TADP DRP implementation. Hence, the GBRT prediction results for wind speed, solar irradiance and the consequent WT and PV powers are validated using error metrics (MAE, RSME and r^2) in order to determine the suitable forecasting condition based on the learning rates α_{gbr}. The total data set contained 17,520 data points with an hourly resolution; from which 75% of the data are adopted for training, and 25% are adopted for testing. Table 5 summarizes the forecasting results based on MAE, RSME and r^2 for the GBRT forecasting model under three α_{gbr} values i.e., α_{gbr} = 0.1, 0.3, 0.5. As it can be noticed, the chosen value of α_{gbr} significantly affects the precision of the GBRT forecasting model.

Table 5. Forecasting results of GBRT model based on MAE, RSME and r^2 considering three α_{gbr} values: $\alpha_{gbr} = \{0.1, 0.3, 0.5\}$.

GBRT Algorithm	Error Metric		$\alpha_{gbr} = 0.1$	$\alpha_{gbr} = 0.3$	$\alpha_{gbr} = 0.5$
	MAE	(m/s)	0.22	0.25	0.28
Wind speed forecasts	RMES	(m/s)	0.27	0.33	0.39
	r^2		0.96	0.94	0.92
	MAE	kW	35.03	39.52	44.33
Wind power forecast	RMES	kW	47.98	55.69	62.83
	r^2		0.96	0.94	0.92
	MAE	W/m^2	15.36	18.55	20.35
Solar irradiance forecast	RMES	W/m^2	29.62	34.87	40.50
	r^2		0.99	0.98	0.98
	MAE	kW	17.43	21.05	23.08
Photovoltaic power forecast	RMES	kW	33.60	39.55	45.94
	r^2		0.99	0.98	0.98

The best wind speed and wind power forecast results are realized when the α_{gbr} value chosen equals 0.1. The least error values indicated by MAE and RMSE of 0.22 (m/s) and 0.27 (m/s) for wind speed prediction and 35.03 kW and 47.98 kW for wind power forecast respectively confirms the consequences of the α_{gbr} value chosen. The results accuracy are further validated using the r^2 metric; the highest value of $r^2 = 0.96$ further establishes that the GBRT at $\alpha_{gbr} = 0.1$ is an appropriate model for wind speed and wind power forecasting. Figure 6. shows a comparison of the actual wind speed versus the predicted wind speed with one-hour-ahead rolling forecasting horizon using the GBRT model when α_{gbr} is set to 0.1 (for the best α_{gbr} value).

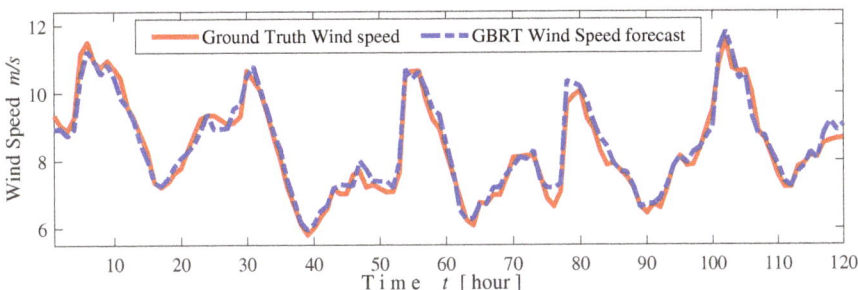

Figure 6. Comparison between the actual versus the predicted wind speed with one-hour-ahead rolling forecasting horizon using the GBRT model at $\alpha_{gbr} = 0.1$ (from 1/12/2018 to 5/12/2018).

Also, for solar irradiance and PV power prediction, the best results are obtained when the α_{gbr} parameter is set to 0.1. The minimum error values indicated by MAE and RMSE of 15.36 (W/m^2) and 29.62 (W/m^2) for solar irradiance prediction and 17.43 kW and 33.60 kW for PV power forecast, respectively, validate the parameter selection. Also, the highest r^2 metric of 0.99 shows the goodness of fit and suitability of the model selection as being appropriate. Figure 7 shows a comparison of the actual versus the predicted solar irradiances with one-hour-ahead rolling forecasting horizon using the GRBT model when α_{gbr} is set to 0.1 (for the best α_{gbr} value).

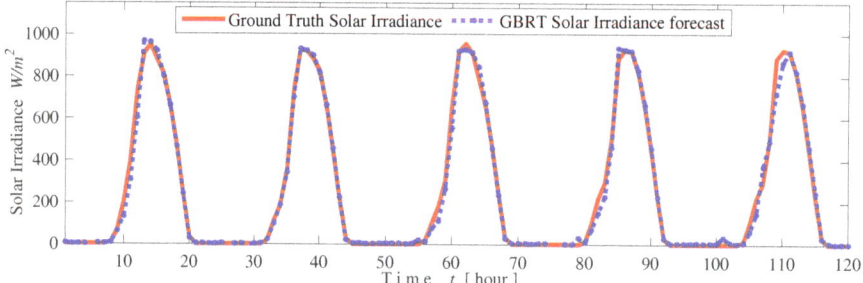

Figure 7. Comparison of the actual versus the predicted solar irradiances with one-hour-ahead rolling forecasting horizon using the GRBT forecasting model at $\alpha_{gbr} = 0.1$ (from 1/12/2018 to 5/12/2018).

Figure 8a,b shows the Pareto front plots with TADP DRP considering BESS and PHES, respectively; and Table 6 summarizes the cost–benefits analysis for the BESS and PHES-based microgrid configuration.

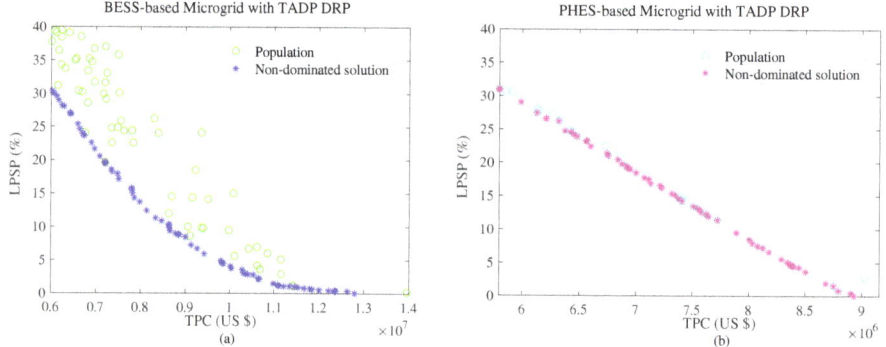

Figure 8. Pareto front plots for case 3.

Table 6. Techno-economic analysis for case 3.

	BESS-Based Microgrid				PHES-Based Microgrid			
LPSP	0%	5%	10%	15%	0%	5%	10%	15%
PV capacity (kW)	1424	1191	1210	1193	1858	1831	1842	1826
WT capacity (kW)	1871	2020	1878	1655	1513	1363	1204	1048
ESS capacity (kWh)	5603	1341	460	181	7494	7326	7121	6397
TPC (US $)	1.28×10^7	9.82×10^6	8.78×10^6	7.83×10^6	8.93×10^6	8.38×10^6	7.89×10^6	7.34×10^6

According to the results of optimal capacity sizing considering TADP DRP at LPSP=0%, it can be noted that adoption PHES-based configuration will results in about 30.23% investment costs reduction compared to the BESS-based. Hence, PHES-based microgrid is the most cost-effective microgrid configuration compared to the BESS-based microgrid design. For all the scenarios (case 1–3) investigated, it is seen that PHES gives the lowest investment cost on ESS compared to BESS. Thus, for a cost-effective long-term investment, it can be deduced that the selection of the PHES-based microgrid has a better economic prospect compared to BESS-based configuration.

6.4. Techno-Economic Comparison for Each ESS Type Based on DRP Options at Maximum System Reliability (LPSP = 0%)

In this section, different microgrid configurations based on the DRP options are evaluated based on the net investment cost for different ESS types. The prospect of each configuration in the long-term microgrid planning with the possibility of high renewable energy fraction is reflected in the net investment cost under each system configuration. Figure 9a,b shows the pareto front plots comparison, without and with DRPs, for BESS and PHES-based microgrid design, respectively.

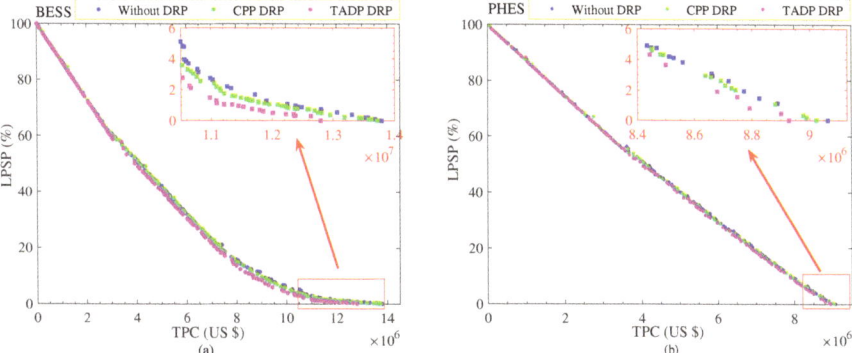

Figure 9. (**a**,**b**) shows the Pareto front plots comparison for BESS and PHES-based microgrid design respectively, based on the DRP flexibility options.

Table 7 summarizes the investment cost under each configuration and flexibility options. The reference cases are the ones without DRP consideration (case 1) and the cost implication of introducing different types of DRP for each ESS-type microgrid are duly analyzed in terms of percentage cost reduction.

Table 7. Techno-economic analysis for each ESS type based on the DRP flexibility options.

	BESS-Based Microgrid			PHES-Based Microgrid		
DRP type	Case 1: without DRP	Case 2: CPP DRP	Case 3: TADP DRP	Case 1: without DRP	Case 2: CPP DRP	Case 3: TADP DRP
PV capacity (kW)	1422	1110	1424	1699	1826	1858
WT capacity (kW)	1919	2165	1871	1657	1561	1513
ESS capacity (kWh)	6798	6436	5603	7800	7789	7494
TPC (US $)	1.38×10^7	1.37×10^7	1.28×10^7	9.06×10^6	9.02×10^6	8.93×10^6
% cost saving	-	0.53%	7.20%	-	0.44%	1.48%

For the BESS-based microgrid, introducing CPP DRP results in a cost saving of 0.53% of the investment from US $ 1.38×10^7 (without DRP) to 1.37×10^7; this cost saving is as a result of 21.59% and 5.33% reduction in the PV and BESS component sizes, respectively. This is because CPP DRP decreased the load demand and consequentially, the BESS dependency during the peak demand periods. For the PHES-based microgrid, the introduction of CPP DRP results in 0.44% cost reduction from US $ 9.06×10^6 to US $ 9.02×10^6. The cost-benefit is because of the 5.8% and 0.14% capacity size reduction of WT and PHES, respectively, and an increase of 7.4% PV capacity. For the two cases, It should be noted that there is a decrease in the investment costs as the CPP DRP shifts the FDR to off-peak from the peak period of the system and ensure a more flattened load profile and prevent sub-optimal capacity sizing.

The potential superiority of TADP DRP over CPP DRP for microgrid design for high renewable energy penetration can be seen in the cost–benefits illustrated in Table 7. The inclusion of the TADP DRP in the BESS-based system resulted in 7.2% cost saving in the total planning costs. The planning cost reduction is due to a decrease of 17.58% and 2.5% for BESS and WT respectively with a slight increase of 0.11% in PV component size. Similarly, this trend is noted for PHES-based system with a total cost reduction of 1.48% resulting from 3.98% and 8.69% decrease in PHES and WT capacities, respectively. However, this results in an increase PV capacity of about 9.36%. Figure 10 illustrates the role and impact of DRPs in minimizing the gap between total generated RES power and load demand profiles. The prospects of TADP DRP over CPP DRP to reduce the mismatch between the load and the RES generated power profiles has been vividly portrayed by significant optimum component size reduction and hence the TPC minimization to realize the techno-economic benefit of a microgrid.

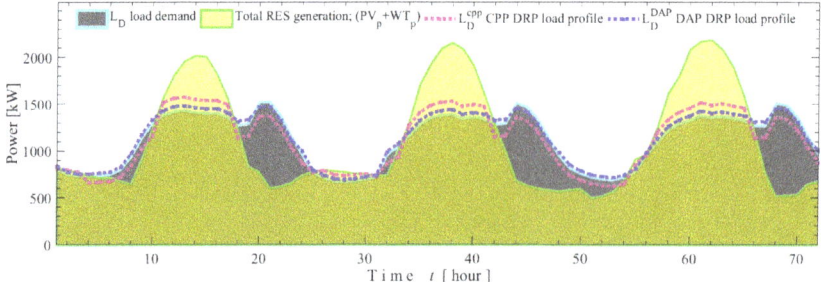

Figure 10. Role and impact of DRPs in minimizing the gap between total generated VRES power and load demand profiles.

Therefore, for the two system and types of DRP investigated, it can be inferred that the application of TADP DRP is more investment-worthy compared to the CPP DRP. TADP DRP short-term flexibility option takes into account the varying generation profile of WT and PV from the forecasting results; thus, the reason for its robustness.

7. Conclusions

This paper investigated the prospects of interlinking the cost of short-term flexibility management of microgrid with the long-term optimal capacity planning models towards achieving a 100% green microgrid by using DRP and forecasting. The long-term capacity planning of energy systems involves the evaluation of the optimal size of each of the system component while the short-term flexibility options are implemented within the optimal energy management strategies. The DRPs are incorporated as flexibility options to minimize the gap between demand and supply, thus minimizing the overall system costs. The forecasting provides an outlook of anticipated generated power proper scheduling for the effective implementation of one of the DRPs employed in this work. The suggested methodology, in this work, seeks to provide a sustainable and cost-effective transformative approach towards achieving a 100% renewable energy generation for Marsabit county microgrid at a reduced cost of investment by cutting down on excessive sizing of system components. This can serve as a benchmark for other under-served isolated regions all over the world.

For the interlinked multi-objective optimization procedure, credible scenarios were investigated considering two ESS technology-based configurations without and with the inclusion of the DRP. DRPs were applied to provide the required operational flexibility that involves shifting the operation of the FDRs from one period to another to minimize the gap between the generation and demand profiles. The two objectives of the techno-economic optimization procedure are the minimization of loss of load probability (LPSP), which is the system reliability criterion and the minimization of the net present value of the investment costs, which is the economic criterion. The forecasting for TADP DRP implementation was performed using the GBRT algorithm on scikit-learn in Python due to its precision and less computational requirement compared to other algorithms, and the MOPSO was adopted for the optimization procedure. The LPSP is set as the standard for economic comparison under each scenario considered in this work. At LPSP = 0%, i.e., maximum system reliability, the potential benefit of TADP DRP outperformed the CPP DRP as reflected on the investment cost component. Also, for the ESS-type performance comparison, PHES was shown to be more cost-effective compared to

BESS due to its low cost per kWh of storage capacity and its resultant economic effect on the whole system configuration.

Author Contributions: conceptualization, M.K.K.; methodology, M.K.K., O.B.A. and M.E.L.; validation, T.S., P.M. and M.A.-A.; formal analysis and investigation, M.K.K., O.B.A. and M.E.L.; writing–original draft preparation, M.K.K. and O.B.A.; writing–review and editing, M.K.K., O.B.A. and M.E.L.; supervision, project administration and funding acquisition, T.S.

Funding: This research received no external funding.

Acknowledgments: The authors wish to acknowledge the Japan international cooperation agency (JICA) for the support provided in the form of African business education (ABE) scholarship to the main author towards the success of this research work. The authors also wish to appreciate the effort of Hannington Gochi of REA, Kenya office for supplying the principal data needed for this project.

Conflicts of Interest: The authors declare no conflict of interest.

Abbreviations

LCC	Life cycle costs.
r	discount rate (%).
REA	Rural electrification authority.
ABE	African business education.
$JICA$	Japan international cooperation agency.
DSM	Demand side management.
DRP	Demand response program.
$PHES$	Pumped Heat energy storage.
ESS	Energy storage system.
PV	Photovoltaic system.
WT	Wind turbine.
$VRES$	Variable renewable energy sources.
$BESS$	Battery energy storage system.
d_r	hourly self-discharge rate of ESS.
CPP	Critical Peak Pricing.
$TADP$	Time-ahead dynamic pricing.
FDR	Flexible demand resources.
FDR^{min}	Minimum FDR limit.
FDR^{max}	Maximum FDR limit.
L_D	Load demand (kW).
L_D^{cpp}	CPP load demand (kW).
L_D^{TADP}	TADP load demand (kW).
E_{std}^{pr}	Standard electricity price.
E_{pr}^{TADP}	TADP electricity price.
$LPSP$	Loss of power supply probability.
PV_p	instantaneous power output of Photovoltaic system (kW).
WT_p	instantaneous power output of Wind turbine (kW).
SOC_{min}	Minimum limit of the SOC.
SOC_{max}	Maximum limit of the SOC.
SOC	State of charge of ESS.
N	Project lifetime
n	year index
T	Total number of time periods, i.e., in a year scheduling horizon
t	instantaneous time index in the scheduling horizon
PV_p^{rtd}	installed rated power of PV (kW).
WT_p^{rtd}	installed rated power of WT (kW).
G	incident solar irradiance (W/m^2).
t_{pv}	temperature of PV module.
θ_t	Temperature coefficient of the PV.

f_{pv} Power reduction factor of PV (%).
G_{std} standard test condition incident solar irradiance (1000 W/m^2).
M_r mass of PHES storage medium.
SH_r specific heat densities of storage medium.
β_{ch} charging efficiency of the ESS.
β_{ds} discharging efficiency of the ESS.
β_c power converters efficiency.
pd incentive payment.
ps penalty payment.

References

1. Senshaw, D.A.; Kim, J.W. Meeting conditional targets in nationally determined contributions of developing countries: Renewable energy targets and required investment of GGGI member and partner countries. *Energy Policy* **2018**, *116*, 433–443. [CrossRef]
2. Hansen, K.; Mathiesen, B.V.; Skov, I.R. Full energy system transition towards 100% renewable energy in Germany in 2050. *Renew. Sustain. Energy Rev.* **2019**, *102*, 1–13. [CrossRef]
3. Bramstoft, R.; Skytte, K. Decarbonizing Sweden's energy and transportation system by 2050. *Int. J. Sustain. Energy Plan. Manag.* **2017**, *14*, 3–20.
4. Borland, J.; Tanaka, T. Overcoming Barriers to 100% Clean Energy for Hawaii Starts at the Bottom of the Energy Food Chain with Residential Island Nano-Grid and Everyday Lifestyle Behavioral Changes. In Proceedings of the 2018 IEEE 7th World Conference on Photovoltaic Energy Conversion (WCPEC) (A Joint Conference of 45th IEEE PVSC, 28th PVSEC 34th EU PVSEC), Waikoloa, WI, USA, 10–15 June 2018; pp. 3829–3834. [CrossRef]
5. Mollenhauer, E.; Christidis, A.; Tsatsaronis, G. Increasing the Flexibility of Combined Heat and Power Plants With Heat Pumps and Thermal Energy Storage. *J. Energy Resour. Technol.* **2018**, *140*, 020907. [CrossRef]
6. Sakah, M.; Diawuo, F.A.; Katzenbach, R.; Gyamfi, S. Towards a sustainable electrification in Ghana: A review of renewable energy deployment policies. *Renew. Sustain. Energy Rev.* **2017**, *79*, 544–557. [CrossRef]
7. Khoodaruth, A.; Oree, V.; Elahee, M.; Clark, W.W. Exploring options for a 100% renewable energy system in Mauritius by 2050. *Utilities Policy* **2017**, *44*, 38–49. [CrossRef]
8. Adewuyi, O.B.; Lotfy, M.E.; Akinloye, B.O.; Howlader, H.O.R.; Senjyu, T.; Narayanan, K. Security-constrained optimal utility-scale solar PV investment planning for weak grids: Short reviews and techno-economic analysis. *Appl. Energy* **2019**, *245*, 16–30. [CrossRef]
9. Aliyu, A.K.; Modu, B.; Tan, C.W. A review of renewable energy development in Africa: A focus in South Africa, Egypt and Nigeria. *Renew. Sustain. Energy Rev.* **2018**, *81*, 2502–2518. [CrossRef]
10. Huang, Y.W.; Kittner, N.; Kammen, D.M. ASEAN grid flexibility: Preparedness for grid integration of renewable energy. *Energy Policy* **2019**, *128*, 711–726. [CrossRef]
11. Papaefthymiou, G.; Dragoon, K. Towards 100% renewable energy systems: Uncapping power system flexibility. *Energy Policy* **2016**, *92*, 69–82. [CrossRef]
12. Hussain, M.; Gao, Y. A review of demand response in an efficient smart grid environment. *Electr. J.* **2018**, *31*, 55–63. [CrossRef]
13. Gong, H.; Wang, H. Day-ahead generation scheduling for variable energy resources considering demand response. In Proceedings of the 2016 IEEE PES Asia-Pacific Power and Energy Engineering Conference (APPEEC), Xi'an, China, 25–28 October 2016; pp. 2076–2080. [CrossRef]
14. Ding, Y.; Shao, C.; Yan, J.; Song, Y.; Zhang, C.; Guo, C. Economical flexibility options for integrating fluctuating wind energy in power systems: The case of China. *Appl. Energy* **2018**, *228*, 426–436. [CrossRef]
15. Taibi, E.; Nikolakakis, T.; Gutierrez, L.; Fernandez, C.; Kiviluoma, J.; Rissanen, S.; Lindroos, T.J. *Power System Flexibility for the Energy Transition: Part 1, Overview for Policy Makers*; International Renewable Energy Agency: Abu Dhabi, UAE, 2018.
16. Ma, J.; Silva, V.; Belhomme, R.; Kirschen, D.S.; Ochoa, L.F. Evaluating and planning flexibility in sustainable power systems. In Proceedings of the 2013 IEEE Power & Energy Society General Meeting, Vancouver, BC, Canada, 21–25 July 2013; pp. 1–11.

17. Zhou, B.; Xu, D.; Li, C.; Chung, C.Y.; Cao, Y.; Chan, K.W.; Wu, Q. Optimal Scheduling of Biogas–Solar–Wind Renewable Portfolio for Multicarrier Energy Supplies. *IEEE Trans. Power Syst.* **2018**, *33*, 6229–6239. [CrossRef]
18. Taibi, E.; Nikolakakis, T.; Gutierrez, L.; Fernandez, C.; Kiviluoma, J.; Rissanen, S.; Lindroos, T.J. *Power System Flexibility for the Energy Transition: Part 2, IRENA FlexTool Methodology*; International Renewable Energy Agency: Abu Dhabi, UAE, 2018.
19. Ralon, P.; Taylor, M.; Ilas, A.; Diaz-Bone, H.; Kairies, K. *Electricity Storage and Renewables: Costs and Markets to 2030*; International Renewable Energy Agency: Abu Dhabi, UAE, 2017.
20. Awan, A.B.; Zubair, M.; Sidhu, G.A.S.; Bhatti, A.R.; Abo-Khalil, A.G. Performance analysis of various hybrid renewable energy systems using battery, hydrogen, and pumped hydro-based storage units. *Int. J. Energy Res.* **2018**. [CrossRef]
21. Zhang, W.; Maleki, A.; Rosen, M.A.; Liu, J. Optimization with a simulated annealing algorithm of a hybrid system for renewable energy including battery and hydrogen storage. *Energy* **2018**, *163*, 191–207. [CrossRef]
22. Khiareddine, A.; Salah, C.B.; Rekioua, D.; Mimouni, M.F. Sizing methodology for hybrid photovoltaic/wind/hydrogen/battery integrated to energy management strategy for pumping system. *Energy* **2018**, *153*, 743–762. [CrossRef]
23. Huang, Y.; Keatley, P.; Chen, H.; Zhang, X.; Rolfe, A.; Hewitt, N. Techno-economic study of compressed air energy storage systems for the grid integration of wind power. *Int. J. Energy Res.* **2018**, *42*, 559–569. [CrossRef]
24. Amrollahi, M.H.; Bathaee, S.M.T. Techno-economic optimization of hybrid photovoltaic/wind generation together with energy storage system in a stand-alone micro-grid subjected to demand response. *Appl. Energy* **2017**, *202*, 66–77. [CrossRef]
25. Jabir, H.; Teh, J.; Ishak, D.; Abunima, H. Impacts of demand-side management on electrical power systems: A review. *Energies* **2018**, *11*, 1050. [CrossRef]
26. Söder, L.; Lund, P.D.; Koduvere, H.; Bolkesjø, T.F.; Rossebø, G.H.; Rosenlund-Soysal, E.; Skytte, K.; Katz, J.; Blumberga, D. A review of demand side flexibility potential in Northern Europe. *Renew. Sustain. Energy Rev.* **2018**, *91*, 654–664. [CrossRef]
27. Chen, Y.; Xu, P.; Gu, J.; Schmidt, F.; Li, W. Measures to improve energy demand flexibility in buildings for demand response (DR): A review. *Energy Build.* **2018**, *177*, 125–139. [CrossRef]
28. Siano, P. Demand response and smart grids—A survey. *Renew. Sustain. Energy Rev.* **2014**, *30*, 461–478. [CrossRef]
29. Hussain, I.; Mohsin, S.; Basit, A.; Khan, Z.A.; Qasim, U.; Javaid, N. A review on demand response: Pricing, optimization, and appliance scheduling. *Procedia Comput. Sci.* **2015**, *52*, 843–850. [CrossRef]
30. Neupane, B.; Pedersen, T.B.; Thiesson, B. Utilizing device-level demand forecasting for flexibility markets. In Proceedings of the Ninth International Conference on Future Energy Systems, Karlsruhe, Germany, 12–15 June 2018; ACM: New York, NY, USA, 2018; pp. 108–118.
31. González-Aparicio, I.; Zucker, A. Impact of wind power uncertainty forecasting on the market integration of wind energy in Spain. *Appl. Energy* **2015**, *159*, 334–349. [CrossRef]
32. Martinez-Anido, C.B.; Botor, B.; Florita, A.R.; Draxl, C.; Lu, S.; Hamann, H.F.; Hodge, B.M. The value of day-ahead solar power forecasting improvement. *Sol. Energy* **2016**, *129*, 192–203. [CrossRef]
33. Notton, G.; Nivet, M.L.; Voyant, C.; Paoli, C.; Darras, C.; Motte, F.; Fouilloy, A. Intermittent and stochastic character of renewable energy sources: Consequences, cost of intermittence and benefit of forecasting. *Renew. Sustain. Energy Rev.* **2018**, *87*, 96–105. [CrossRef]
34. Gürtler, M.; Paulsen, T. The effect of wind and solar power forecasts on day-ahead and intraday electricity prices in Germany. *Energy Econ.* **2018**, *75*, 150–162. [CrossRef]
35. Xydas, E.; Qadrdan, M.; Marmaras, C.; Cipcigan, L.; Jenkins, N.; Ameli, H. Probabilistic wind power forecasting and its application in the scheduling of gas-fired generators. *Appl. Energy* **2017**, *192*, 382–394. [CrossRef]
36. Wang, Q.; Wu, H.; Florita, A.R.; Martinez-Anido, C.B.; Hodge, B.M. The value of improved wind power forecasting: Grid flexibility quantification, ramp capability analysis, and impacts of electricity market operation timescales. *Appl. Energy* **2016**, *184*, 696–713. [CrossRef]
37. Guangqian, D.; Bekhrad, K.; Azarikhah, P.; Maleki, A. A hybrid algorithm based optimization on modeling of grid independent biodiesel-based hybrid solar/wind systems. *Renew. Energy* **2018**, *122*, 551–560. [CrossRef]

38. Ramli, M.A.; Bouchekara, H.; Alghamdi, A.S. Optimal sizing of PV/wind/diesel hybrid microgrid system using multi-objective self-adaptive differential evolution algorithm. *Renew. Energy* **2018**, *121*, 400–411. [CrossRef]
39. Nadjemi, O.; Nacer, T.; Hamidat, A.; Salhi, H. Optimal hybrid PV/wind energy system sizing: Application of cuckoo search algorithm for Algerian dairy farms. *Renew. Sustain. Energy Rev.* **2017**, *70*, 1352–1365. [CrossRef]
40. Azaza, M.; Wallin, F. Multi objective particle swarm optimization of hybrid micro-grid system: A case study in Sweden. *Energy* **2017**, *123*, 108–118. [CrossRef]
41. Gazijahani, F.S.; Salehi, J. Reliability constrained two-stage optimization of multiple renewable-based microgrids incorporating critical energy peak pricing demand response program using robust optimization approach. *Energy* **2018**, *161*, 999–1015. [CrossRef]
42. Kharrich, M.; Akherraz, M.; Sayouti, Y. Optimal sizing and cost of a Microgrid based in PV, WIND and BESS for a School of Engineering. In Proceedings of the 2017 International Conference on Wireless Technologies, Embedded and Intelligent Systems (WITS), Fez, Morocco, 19–20 April 2017; pp. 1–5.
43. Balali, M.H.; Nouri, N.; Rashidi, M.; Nasiri, A.; Otieno, W. A multi-predictor model to estimate solar and wind energy generations. *Int. J. Energy Res.* **2018**, *42*, 696–706. [CrossRef]
44. Howes, J. Concept and development of a pumped heat electricity storage device. *Proc. IEEE* **2012**, *100*, 493–503. [CrossRef]
45. Desrues, T.; Ruer, J.; Marty, P.; Fourmigué, J. A thermal energy storage process for large scale electric applications. *Appl. Therm. Eng.* **2010**, *30*, 425–432. [CrossRef]
46. Energy Storage Association. Pumped Heat Electrical Storage (PHES). 2019. Available online: http://energystorage.org/energy-storage/technologies/pumped-heat-electrical-storage-phes (accessed on 15 October 2018).
47. White, A.; Parks, G.; Markides, C.N. Thermodynamic analysis of pumped thermal electricity storage. *Appl. Therm. Eng.* **2013**, *53*, 291–298. [CrossRef]
48. McTigue, J.D.; White, A.J.; Markides, C.N. Parametric studies and optimisation of pumped thermal electricity storage. *Appl. Energy* **2015**, *137*, 800–811. [CrossRef]
49. Conteh, A.; Lotfy, M.E.; Kipngetich, K.M.; Senjyu, T.; Mandal, P.; Chakraborty, S. An Economic Analysis of Demand Side Management Considering Interruptible Load and Renewable Energy Integration: A Case Study of Freetown Sierra Leone. *Sustainability* **2019**, *11*, 2828. [CrossRef]
50. Aalami, H.; Moghaddam, M.P.; Yousefi, G. Evaluation of nonlinear models for time-based rates demand response programs. *Int. J. Electr. Power Energy Syst.* **2015**, *65*, 282–290. [CrossRef]
51. Javaid, N.; Ahmed, A.; Iqbal, S.; Ashraf, M. Day ahead real time pricing and critical peak pricing based power scheduling for smart homes with different duty cycles. *Energies* **2018**, *11*, 1464. [CrossRef]
52. Aalami, H.; Moghaddam, M.P.; Yousefi, G. Modeling and prioritizing demand response programs in power markets. *Electr. Power Syst. Res.* **2010**, *80*, 426–435. [CrossRef]
53. Persson, C.; Bacher, P.; Shiga, T.; Madsen, H. Multi-site solar power forecasting using gradient boosted regression trees. *Sol. Energy* **2017**, *150*, 423–436. doi:10.1016/j.solener.2017.04.066. [CrossRef]
54. Coello, C.A.C.; Pulido, G.T.; Lechuga, M.S. Handling multiple objectives with particle swarm optimization. *IEEE Trans. Evol. Comput.* **2004**, *8*, 256–279. [CrossRef]
55. Weather History Download. Available online: https://www.meteoblue.com (accessed on 11 January 2019).
56. Photovoltaic Geographical Information System. Available online: https://rem.jrc.ec.europa.eu (accessed on 13 January 2019).
57. *Power Generation and Transmission Master Plan, Kenya Medium Term Plan 2015–2020—Vol. I*; Energy and Petroleum Regulatory Authority: Nairobi, Kenya, 2018.
58. Tarrif Setting: Electricity. Available online: https://www.erc.go.ke/services/economic-regulation (accessed on 14 January 2019).
59. Electricity Cost in Kenya. Available online: https://stima.regulusweb.com/ (accessed on 11 December 2018).

60. Benato, A. Performance and cost evaluation of an innovative Pumped Thermal Electricity Storage power system. *Energy* **2017**, *138*, 419–436. [CrossRef]
61. Smallbone, A.; Jülch, V.; Wardle, R.; Roskilly, A.P. Levelised Cost of Storage for Pumped Heat Energy Storage in comparison with other energy storage technologies. *Energy Convers. Manag.* **2017**, *152*, 221–228. [CrossRef]

© 2019 by the authors. Licensee MDPI, Basel, Switzerland. This article is an open access article distributed under the terms and conditions of the Creative Commons Attribution (CC BY) license (http://creativecommons.org/licenses/by/4.0/).

Article

Can Large Educational Institutes Become Free from Grid Systems? Determination of Hybrid Renewable Energy Systems in Thailand

Eunil Park [1], Sang Jib Kwon [2,*] and Angel P. del Pobil [1,3,*]

1 Department of Interaction Science, Sungkyunkwan University, Seoul 03063, Korea; eunilpark@skku.edu
2 Department of Business Administration, Dongguk University, Gyeongju 38066, Korea
3 Computer Science and Engineering Department, Jaume-I University, 12071 Castellon, Spain
* Correspondence: risktaker@dongguk.ac.kr (S.J.K.); pobil@uji.es (A.P.d.P.);
   Tel.: +82-54-770-2324 (S.J.K.); +34-964-72-82-93 (A.P.d.P.)

Received: 6 May 2019; Accepted: 30 May 2019; Published: 5 June 2019

Abstract: In some countries, renewable energy resources have become one of the mainstreams of energy savings and sustainable development. Thailand is one of the major countries to use renewable energy generation facilities in public buildings. In particular, public educational institutes consume large amounts of electricity from the grid. To reduce the electricity dependency on the national grid connection and greenhouse gas emissions, this paper introduces potential optimized solutions of renewable energy generation systems for a public university in Thailand, Chiang Mai University. Based on the simulation results from HOMER software, the potential configuration organized by PV panels, batteries and converters is proposed. The suggested configuration achieves 100% of the renewable fraction with $0.728 of the cost of energy for per electricity. Moreover, the greenhouse gas emissions are significantly reduced. Both the implications and limitations are presented based on simulation results.

Keywords: renewable energy; sustainability; greenhouse gas emission; economic feasibility

1. Introduction

Owing to the considerable social and environmental concerns, environment and energy issues are two of the main motivations of global sustainable development [1]. In particular, certain countries have struggled to achieve two goals, economic growth and energy savings [2]. Among these countries, Thailand is one of the major countries attempting to contribute energy savings [3]. In 2013, approximately 8.58% of the final electricity consumption was produced by total renewable energy resources (14,107 GWh from 164,322 GWh of the final consumption) [4]. Although this share is not insignificant compared to other countries, the electricity generated from renewable energy resources, which are one of the most appropriate to use renewable energy facilities, could be larger than the current amount of renewable energy facilities [5]. Moreover, the majority of renewable energy facilities currently used in Thailand are hydro and biomass facilities (Table 1; [5]). Therefore, solar and wind energy have significant potential.

Moreover, because Thailand which is one of the nations in the United Nations Framework Convention on Climate Change (UNFCCC), agreed the Paris Agreement which presents the Intended Nationally Determined Contribution (INDC), the government of Thailand should attempt to reduce the emission of greenhouse gases (GHG) by utilizing renewable energy resources [6]. Table 2 summarizes key descriptions which are applied to Thailand.

Table 1. Current status of electricity production from renewable energy facilities in Thailand [5].

Sources	Amount (GWh)	Share
Primary solid biofuels	6141	43.50%
Hydro	5748	40.70%
Solar PV	1080	7.70%
Biogases	539	3.80%
Wind	305	2.20%
Municipal waste	293	2.10%
Geothermal	1	-
Solar thermal	<1.0	-
Tide, wave, ocean	<1.0	-

Table 2. Key points which are applied to Thailand in the Paris Agreement [6,7].

Item	Descriptions
Greenhouse gas emissions	20% reduction of GHG emissions compared to the projected BAU (business-as-usual) target in 2030
Global average temperature increase	Below 2 celcius degrees
National renewable energy targets to respond the Paris agreement	30% of total energy consumption from renewable energy resources in 2036

As the initial part of Thailand national government's contribution, the government has aimed to apply renewable energy facilities in public buildings for energy savings [8]. Among these buildings, public education institutes are required to contribute to energy saving through the installation of renewable and sustainable energy facilities [9].

Currently, Thailand has employed a long-term national energy and electricity planning policy which is called as the Power Development Plan (PDP) from 2015 to 2036 [10]. The majority of PDP considers the production and distribution of renewable energy facilities in Thailand. That is, renewable energy and its facilities are among the top priorities in the successful applications of PDP. Because dependence on fossil fuels can be environmentally and economically unsustainable with notable heavy burdens on the national economy, Thailand's government hopes to fully revise its national energy systems with renewable energy. Based on the key concept of PDP, the Alternative Energy Development Plan 2015 was introduced and employed for the reduction of dependence on fossil fuels and the promotion of using alternative energy facilities from 7279 MW to 19,635 MW-capacity (2014–2036).

However, only few studies have investigated and explored the potentiality and possibilities of renewable energy facilities in Southeast Asia. Table 3 summaries the findings of previous studies which were conducted in Southeast Asia.

Table 3. Examples of the suggested configuration of renewable energy production systems in Southeast Asia.

Regions	Year	Configuration	Cost and Renewable Fraction
Maldives [11]	2018	PV-diesel generator-battery	$0.245 per kWh of COE (cost of energy) and 30% of renewable fraction
Indonesia [12]	2013	PV-wind turbine-battery	$0.751 per kWh of COE and 100% of renewable fraction
Myanmar [13]	2018	PV-diesel generator-battery	$0.193–$1.830 per kWh of COE
Thailand [14]	2002	PV-diesel generator-battery	$0.589 per kWh of COE and 36.9% of renewable fraction
Cambodia [15]	2017	PV-diesel generator-battery	$0.377 per kWh of COE and 13.0% of renewable fraction
Malaysia [16]	2017	PV-battery	$1.220 per kWh of COE and 100% of renewable fraction

As presented in Table 3 and the findings of previous studies conducted in Southeast Asia, there are notable economic burdens in successfully diffusing renewable energy production facilities. Thus, several nations have attempted to preferentially employ the facilities with the considerations of their public institutions and organizations [17,18].

Therefore, the current study introduces the optimal configuration of renewable energy generation systems for Chiang Mai University, which is one of the largest public universities in Thailand. Using HOMER software (Hybrid Renewable and Distributed Generation System), the possible components of the configuration are introduced by reducing the environmental pollution and the dependence on the national grid system. Although there are notable limitations of HOMER software in exploring the feasibility of renewable resources including the needs of time-series datasets, notable time consumption, and certain criteria on converge, HOMER software can consider multiple combinations of different energy-related technologies, provide relatively precise results, and present optimized configurations of energy production systems [19]. That is, the current study aims to respond to the following research questions.

- **Research question 1** What is the optimal renewable electricity production system for Chiang Mai University in Thailand?
- **Research question 2** How much are the amount of greenhouse gas emissions reduced by as a result of using the optimal system for the university?

2. Chiang Mai University

2.1. Location and Facilities

Chiang Mai University is one of the largest universities in Thailand [20]. Because the university is public, "*the Energy Conservation Promotion Act of Thailand for government building*" should be applied [21]. This means that energy conservation and saving facilities should be constructed for the buildings. Under the act, the establishment of these facilities is fully supported by the government. In the university, there are approximately 170 buildings. Although the university is organized in four separate campuses, the main campus, Suan Sak Campus, has the main electricity demand of the university. The latitude and longitude of the university are 18.80° N and 98.95° E, respectively. This means that the main campus is located approximately 5 km-west from the center of the city. In 2015, approximately 36,000 students and 2500 staff worked in the university. Figure 1 shows the location of Chiang Mai University, Thailand [20].

Figure 1. The location of Chiang Mai University (created by the authors).

2.2. Load Information

The current electricity system of Chiang Mai University is operated by the national grid system. The amount of electricity consumed in 2015 was calculated to be 17,654,195 kWh. Because this amount is too heavy to simulate, the current study used the 50% scaled electricity load information for the simulation. Based on the 50% scaled electricity load information, 1385 kW of peak electricity and 19,472 kWh/d of average daily peak electricity were examined. The load factor in 2015 was calculated to be 0.586 (Figure 2).

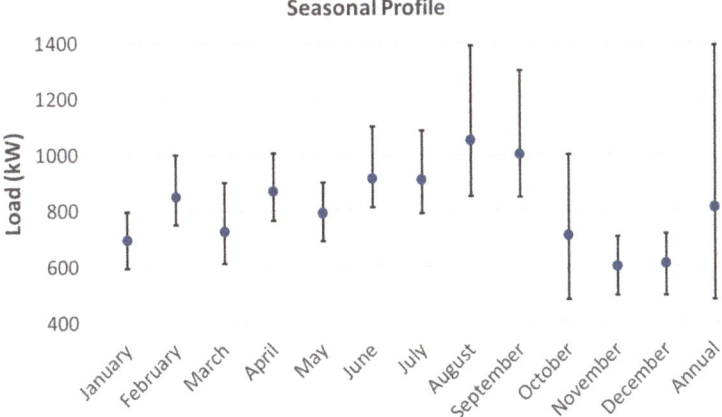

Figure 2. Monthly seasonal electricity load profile of Chiang Mai University (created by the authors).

2.3. Wind Resources

The wind resource datasets of Chiang Mai University were obtained from the Thai Meteorological Department (2014) [22]. Because the height of wind turbines currently operated in Thailand is 25 m, the wind speed at 25 m was considered to be intermediate between that at 50 m and at the ground. Figure 3 shows the monthly average wind speed of the university. The annual average wind speed is 2.507 m/s.

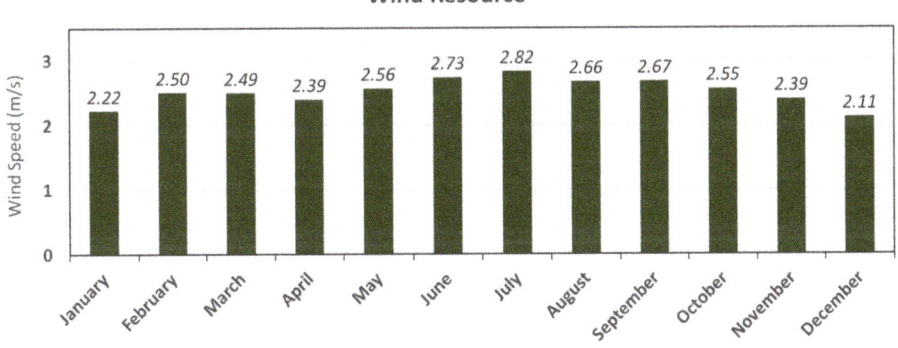

Figure 3. Wind resource information of Chiang Mai University (created by the authors).

2.4. Solar Resources

The datasets provided by the National Aeronautics and Space Administration (NASA) were used as the information of solar resources in the simulation [23]. Figure 4 presents the annual baseline datasets of the solar resources. Based on the datasets, 0.554 of the annual solar clearness index and 5.257 kWh/m^2/d of the solar average daily radiation are presented. The definition of solar clearness index is defined as *"the ratio of the daily horizontal radiation to the extraterrestrial value"* [24].

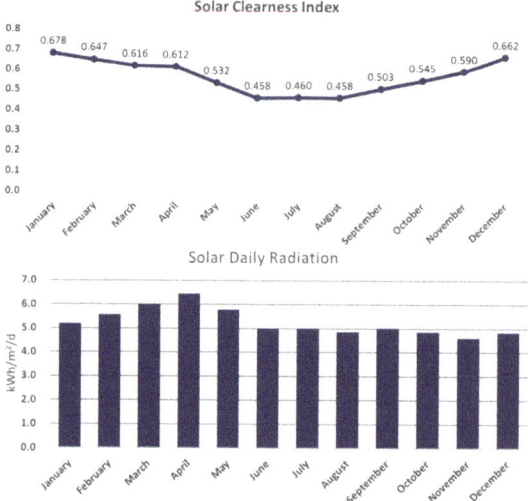

Figure 4. Solar resource information of Chiang Mai University (created by the authors).

3. Key Parameters for the Simulation

3.1. Annual Real Interest Rate

To calculate the accurate economic results from the simulation, the annual real interest rate in Thailand should be input in the HOMER simulations [25]. Based on the official introduction of the World Bank, an annual real interest rate of 5.38% was used [26].

3.2. Economic Evaluations

Before the considerations of economic evaluations, the current study only considers the configurations which can achieve 100% of renewable fraction. To evaluate the simulation results, the optimal configurations were ranked by two economic outputs, the cost of energy (COE) and the net present cost (NPC). The COE is referred to as *"the average consumed cost in producing 1 kWh from the suggested system"* [27]. Moreover, the NPC is *"the consumed cost in establishing, operating, maintaining, and replacing the components of the suggested system in the project lifetime"* [28,29]. Based on a previous simulation background, the project lifetime was assumed to be 25 years. Other specific economic methods and calculations used in the simulations were employed by the validated examinations introduced by [30].

3.3. Environmental Parameters

Based on the electricity and energy generation information of the traditional grid system, 632 g of CO_2 (carbon dioxide), 2.74 g of SO_2 (sulfur dioxide), and 1.34 g of NO and NO_2 (nitrogen oxides) are reduced when the grid system does not need to generate 1 kWh of electricity.

4. Renewable Electricity Generation Systems

To propose independent renewable electricity generation systems, PV arrays, wind turbines, batteries, and a converter were employed as the possible components for organizing the systems. Table 4 lists the cost specifications of the components used in the simulations based on the cost information of the components in prior studies [27–30]. HOMER was used to present the optimal configurations of possible renewable electricity generation systems for Chiang Mai University.

Table 4. Detailed economic and technical information of the components in the simulation [27–30] (O&M cost: Operation & Management cost; created by the authors).

Components	Size	Capital Cost ($)	Replacement Cost ($)	O&M Cost ($ per Year)	Lifetime (Years)	Considered Capacity	Others
PV array	1.0 kW	1800	1800	25	20	0–25,000 kW (5 kW-capacity steps)	• Derating factor: 80% • Ground reflectance: 20% [31]
Wind turbine	2 units	29,000	29,000	400	15	0–100 units (2-unit steps)	• Type: Generic 10 kW turbine • Hub height: 25 m [32]
Battery	1 unit	1229	1229	10	– (hour-oriented)	0–30,000 units (5-unit steps)	• Nominal capacity: 1156 Ah and 6.94 kWh • Round trip efficiency: 80% • Nominal voltage: 6 V • Lifetime throughput: 9645 kWh [33]
Converter	1.0 kW	800	800	10	15	0–2500 kW (5 kW-capacity steps)	• Efficiency: 90%

5. Results

Table 5 and Figure 5 list the optimal configuration composed by PV arrays, wind turbines, a converter, and batteries. Table 6 shows the total and annual costs of the components in the simulation. The combination of 12,780 kW-capacity of the PV arrays, 17,965 battery units with a 1525 kW-capacity of the electric converter is suggested to respond to the electricity demand of Chiang Mai University.

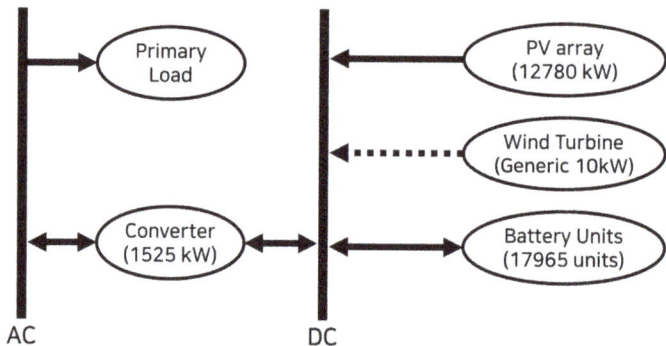

Figure 5. The suggested configuration (created by the authors).

Table 5. Optimal configuration for Chiang Mai University (created by the authors).

Components	Index	Components	Index
PV array	12,780 kW	Initial capital cost	$46,607,984
Wind turbine	0 unit	Operating cost	$1,734,385 per year
Battery	17,965 units	Total net present cost	$70,147,848
Converter	1525 kW	Cost of energy	$0.728 per kWh
Renewable fraction	100%		

Table 6. Total and annual costs of the optimal configuration (created by the authors).

Category	Component	Capital ($)	Replacement ($)	O&M ($)	Salvage ($)	Total ($)
Total cost	PV array	23,004,000	8,065,651	4,336,400	−4,654,895	30,751,156
	Batteries	22,078,984	18,050,268	2,438,292	−5,460,543	37,106,988
	Converter	1,525,000	694,860	206,980	−137,149	2,289,690
	System	46,607,984	26,810,779	6,981,672	−10,252,586	70,147,848
Annual cost	PV array	1,694,904	594,266	319,500	−342,966	2,265,704
	Batteries	1,626,750	1,329,919	179,650	−402,325	2,733,993
	Converter	112,360	51,196	15,250	−10,105	168,701
	System	3,434,013	1,975,382	514,400	−755,397	5,168,399

The optimal configuration shows $5,168,399 of the annual costs with $0.728 per kWh of the COE level. The cash flow is introduced in Figure 6. The annual electricity production was estimated to be 20,768,330 kWh. Figure 7 presents the monthly electricity production. The monthly PV power production and battery state of charge are presented in Figures 8 and 9, respectively.

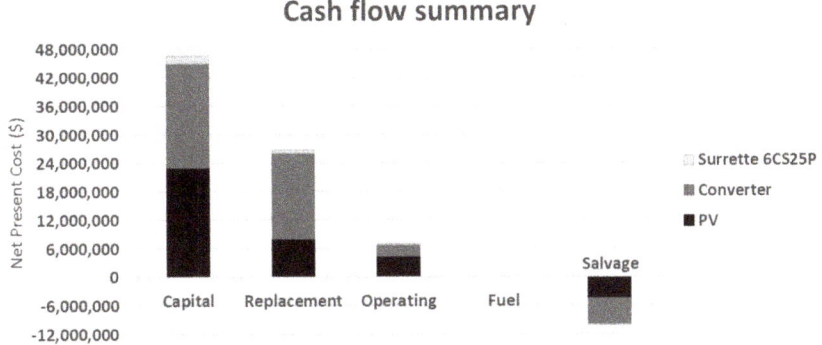

Figure 6. Summary of cash flow in the suggested configuration (created by the authors).

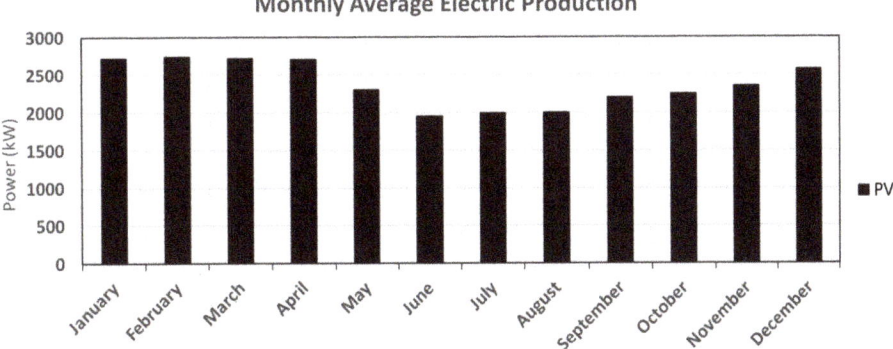

Figure 7. Monthly production of electricity from the suggested configuration (created by the authors).

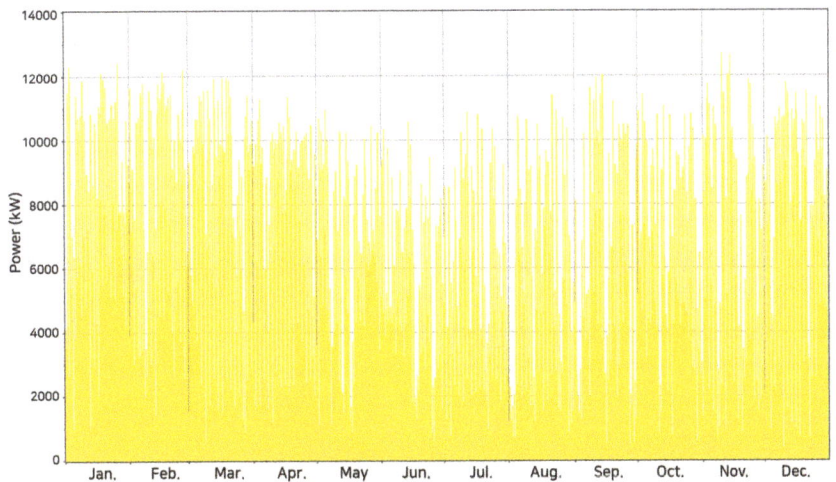

Figure 8. Monthly PV power production of the suggested configuration (created by the authors).

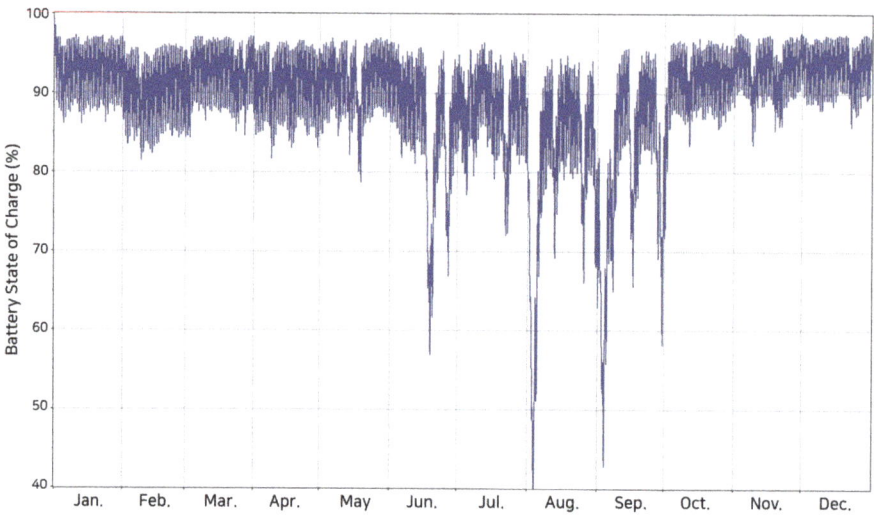

Figure 9. Monthly battery state of charge of the suggested configuration (created by the authors).

The key findings from the simulation results in the current study could be introduced as follows. First, the combination of PV array-batteries-converter was proposed for Chiang Mai University. Second, the suggested configuration from the simulation shows $70,147,828 of the total NPC level with $0.728 kWh of the COE level. Third, the optimal configuration meets the 100% renewable fraction, because the purpose of this study was to present independent renewable electricity generation systems for Chiang Mai University.

Moreover, there are the notable amounts of the annual reduced environmental pollutants of the proposed configurations, instead of using the current grid system. 4,487,738 kg of CO_2, 19,456 kg of SO_2, and 9515 kg of NO and NO_2 cannot be annually emitted by employing the proposed configuration in this study.

6. Discussion and Conclusions

To respond rapidly to the increased electricity demand in countries with sustainable development, and to reduce environmental pollution, several countries have set national plans and policies for renewable energy production facilities [34]. Following this effort, the current study proposes the potential configuration of renewable energy production facilities for Chiang Mai University in Thailand to utilize local renewable resources. Two economic evaluations, COE and NPC, were used to assess the economic feasibility of the configuration. Related to research question 1, the potentially optimal configuration was organized by 12,780 kW-capacity PV array, 17,965 battery units, and 1525 kW-capacity electronic converter.

The configuration, which was composed of a PV array, a converter and batteries with a 5.38% annual real interest rate, achieved a $0.728 per kWh COE with a 100% renewable fraction. The results of the simulation shows the possibility of an eco-friendly campus in Thailand by presenting the potential configuration of renewable energy generation systems for Chiang Mai University. Although the simulation results show heavy initial capital costs, the suggested systems can be practical in allowing the university to be a long-term eco-friendly campus. In addition, because the simulations did not consider the national grid system, which is used as the current electricity system of the university, the suggested systems can achieve greater performance by trading the electricity between the suggested systems and the grid connection. Moreover, using the suggested system shows the significantly reduced environmental pollutants. Related to research question 2, the emissions of greenhouse gas are notably reduced. Moreover, compared to the current electricity system of Chiang Mai University, 179,510 kg of CO_2, 778 kg of SO_2, and 381 kg of NO and NO_2 can be annually eliminated when the suggested system is installed and operated. It means that using the suggested system can provide environmental benefits for the university.

Compared to the findings of several previous studies conducted in Southeast Asia [12,14], the simulation results of the current study indicated that the suggested configuration can achieve 100% of renewable fraction with $0.728 per kWh of COE. Considering the suggested configuration of previous studies in Thailand [14], the suggested configuration in the current study excluded the usage of diesel generators. Considering about $0.858 per kWh of COE is provided by the national grid system in Thailand [35], the COE level presented by the suggested system, $0.728 per kWh of COE, is considered as the economical configuration.

This study had several limitations. First, other policies on renewable energy in Thailand were not considered. For example, the Thailand government started to apply feed-in-tariff policies to power production facilities [36,37]. Second, economic theories that can be used in the energy industry were not considered in the simulations. Prior studies found that there are notable economic theories validated in the renewable energy industry [38]. Third, the economic dynamics of developing countries were not considered. Several scholars indicated that the economic dynamics of developing countries can be a main hindrance to diffusing renewable energy facilities [39]. For example, the pay back period with the internal rate of return of the suggested system can be considered. Third, because the amount of electricity considered in Chiang Mai University is significantly heavy to simulate (17,654,195 kWh), the current study employs the 50% scaled electricity load information. Therefore, future studies should extend the findings of the current study by addressing these limitations.

Author Contributions: Conceptualization: E.P. and A.P.d.P.; Methodology: E.P. and S.J.K.; Data Collection and Simulation: E.P. and S.J.K.; Resources and Design: S.J.K.; Writing: E.P. and A.P.d.P.; Validation and Revision: E.P., S.J.K. and A.P.d.P.

Funding: This work was supported by the Ministry of Education of the Republic of Korea and the National Research Foundation of Korea (NRF-2018S1A5A8027730), and by the Dongguk University Research Fund of 2018. In addition, We acknowledge support from Universitat Jaume I (UJI-B2018-74).

Conflicts of Interest: The authors declare no conflict of interest.

References

1. Abdallah, S.M.; Bressers, H.; Clancy, J.S. Energy reforms in the developing world: Sustainable development compromised? *Int. J. Sustain. Energy Plan. Manag.* **2015**, *5*, 41–56.
2. Black, G.; Black, M.A.T.; Solan, D.; Shropshire, D. Carbon free energy development and the role of small modular reactors: A review and decision framework for deployment in developing countries. *Renew. Sustain. Energy Rev.* **2015**, *43*, 83–94. [CrossRef]
3. Hu, J.L.; Kao, C.H. Efficient energy-saving targets for APEC economies. *Energy Policy* **2007**, *35*, 373–382. [CrossRef]
4. International Energy Agency. Global Renewable Energy, IEA/IRENA Joint Policies and Measures Database, Thailand Statistics. 2015. Available online: http://www.iea.org/policiesandmeasures/renewableenergy/?country=Thailand (accessed on 30 December 2017).
5. Kumar, S. Assessment of renewables for energy security and carbon mitigation in Southeast Asia: The case of Indonesia and Thailand. *Appl. Energy* **2016**, *163*, 63–70. [CrossRef]
6. Chaiyapa, W.; Esteban, M.; Kameyama, Y. Why go green? Discourse analysis of motivations for Thailand's oil and gas companies to invest in renewable energy. *Energy Policy* **2018**, *120*, 448–459. [CrossRef]
7. International Energy Agency. Nationally Determined Contribution (NDC) to the Paris Agreement: Thailand. 2015. Available online: https://www.iea.org/policiesandmeasures/pams/thailand/name-155226-en.php (accessed on 28 May 2019).
8. Industrial Efficiency Policy Database. TH-3: Thailand 20-Year Energy Efficiency Development Plan (2011–2030) (EEDP). 2011. Available online: http://iepd.iipnetwork.org/policy/thailand-20-year-energy-efficiency-development-plan-2011-2030-eedp (accessed on 30 December 2017).

9. Ministry of Energy in Thailand. Thailand's Energy Situation. 2011. Available online: http://www.thailandenergyeducation.com/energy-media/energy-in-thailand (accessed on 30 December 2017).
10. Aroonrat, K.; Wongwises, S. Current status and potential of hydro energy in Thailand: A review. *Renew. Sustain. Energy Rev.* **2015**, *46*, 70–78. [CrossRef]
11. Ali, I.; Shafiullah, G.; Urmee, T. A preliminary feasibility of roof-mounted solar PV systems in the Maldives. *Renew. Sustain. Energy Rev.* **2018**, *83*, 18–32. [CrossRef]
12. Hiendro, A.; Kurnianto, R.; Rajagukguk, M.; Simanjuntak, Y.M.; Junaidi. Techno-economic analysis of photovoltaic/wind hybrid system for onshore/remote area in Indonesia. *Energy* **2013**, *59*, 652–657. [CrossRef]
13. Kim, H.; Jung, T.Y. Independent solar photovoltaic with Energy Storage Systems (ESS) for rural electrification in Myanmar. *Renew. Sustain. Energy Rev.* **2018**, *82*, 1187–1194. [CrossRef]
14. Fung, C.C.; Rattanongphisat, W.; Nayar, C. A simulation study on the economic aspects of hybrid energy systems for remote islands in Thailand. In Proceedings of the 2002 IEEE Region 10 Conference on Computers, Communications, Control and Power Engineering, (TENCOM'02), Beijing, China, 28–31 October 2002; Volume 3, pp. 1966–1969.
15. Lao, C.; Chungpaibulpatana, S. Techno-economic analysis of hybrid system for rural electrification in Cambodia. *Energy Procedia* **2017**, *138*, 524–529. [CrossRef]
16. Halabi, L.M.; Mekhilef, S.; Olatomiwa, L.; Hazelton, J. Performance analysis of hybrid PV/diesel/battery system using HOMER: A case study Sabah, Malaysia. *Energy Convers. Manag.* **2017**, *144*, 322–339. [CrossRef]
17. Aunphattanasilp, C. From decentralization to re-nationalization: Energy policy networks and energy agenda setting in Thailand (1987–2017). *Energy Policy* **2018**, *120*, 593–599. [CrossRef]
18. Carfora, A.; Pansini, R.V.; Romano, A.; Scandurra, G. Renewable energy development and green public policies complementarities: The case of developed and developing countries. *Renew. Energy* **2018**, *115*, 741–749. [CrossRef]
19. Okedu, K.E.; Uhunmwangho, R. Optimization of renewable energy efficiency using HOMER. *Int. J. Renew. Energy Res.* **2014**, *4*, 421–427.
20. Chiang Mai University. The Overview of Chiang Mai University. 2016. Available online: http://www.cmu.ac.th/index_eng.php (accessed on 30 December 2017).
21. Thailand Government. The Energy Conservation Promotion Act. 2016. Available online: www.eppo.go.th/images/law/ENG/nation2.pdf (accessed on 30 December 2017).
22. Thai Meteorological Department. Weather Forecast. 2017. Available online: http://www.tmd.go.th/en (accessed on 30 December 2017).
23. National Aeronautics and Space Administration. POWER Project Data Sets. 2018. Available online: https://power.larc.nasa.gov/ (accessed on 30 December 2018).
24. Braun, J.E. Reducing energy costs and peak electrical demand through optimal control of building thermal storage. *ASHRAE Trans.* **1990**, *96*, 876–888.
25. Park, E.; Kwon, S.J. Towards a Sustainable Island: Independent optimal renewable power generation systems at Gadeokdo Island in South Korea. *Sustain. Cities Soc.* **2016**, *23*, 114–118. [CrossRef]
26. YCharts. Thailand Real Interest Rate. 2016. Available online: https://ycharts.com/indicators/thailand_real_interest_rate (accessed on 30 December 2017).
27. Park, E.; Kwon, S.J. Solutions for optimizing renewable power generation systems at Kyung-Hee University's Global Campus, South Korea. *Renew. Sustain. Energy Rev.* **2016**, *58*, 439–449. [CrossRef]
28. Yoo, K.; Park, E.; Kim, H.; Ohm, J.Y.; Yang, T.; Kim, K.J.; Chang, H.J.; del Pobil, A.P. Optimized renewable and sustainable electricity generation systems for Ulleungdo Island in South Korea. *Sustainability* **2014**, *6*, 7883–7893. [CrossRef]
29. Park, E.; Kwon, S.J. Renewable electricity generation systems for electric-powered taxis: The case of Daejeon metropolitan city. *Renew. Sustain. Energy Rev.* **2016**, *58*, 1466–1474. [CrossRef]
30. Dursun, B. Determination of the optimum hybrid renewable power generating systems for Kavakli campus of Kirklareli University, Turkey. *Renew. Sustain. Energy Rev.* **2012**, *16*, 6183–6190. [CrossRef]
31. Hamad, A.A.; Alsaad, M.A. A software application for energy flow simulation of a grid connected photovoltaic system. *Energy Convers. Manag.* **2010**, *51*, 1684–1689. [CrossRef]
32. Acosta, J.L.; Combe, K.; Djokic, S.Ž.; Hernando-Gil, I. Performance assessment of micro and small-scale wind turbines in urban areas. *IEEE Syst. J.* **2012**, *6*, 152–163. [CrossRef]

33. Lau, K.Y.; Yousof, M.; Arshad, S.; Anwari, M.; Yatim, A. Performance analysis of hybrid photovoltaic/diesel energy system under Malaysian conditions. *Energy* **2010**, *35*, 3245–3255. [CrossRef]
34. Omer, A.M. Energy, environment and sustainable development. *Renew. Sustain. Energy Rev.* **2008**, *12*, 2265–2300. [CrossRef]
35. Piyasil, P. Electricity Pricing in the Residential Sector of Thailand. 2015. Available online: http://www.meconproject.com/wp-content/uploads/report/[Task%206-Electricity%20pricing%20in%20the%20residential%20sector]%20Thailand%20country%20report.pdf (accessed on 28 May 2019).
36. International Energy Agency. Feed-in Tariff for Distributes Solar Systems. 2013. Available online: http://www.iea.org/policiesandmeasures/pams/thailand/name-43052-en.php (accessed on 30 December 2017).
37. International Energy Agency. Feed-in Tariff for Very Small Power Producers (VSPP). 2014. Available online: http://www.iea.org/policiesandmeasures/pams/thailand/name-146463-en.php (accessed on 30 December 2017).
38. Hong, S.; Chung, Y.; Woo, C. Scenario analysis for estimating the learning rate of photovoltaic power generation based on learning curve theory in South Korea. *Energy* **2015**, *79*, 80–89. [CrossRef]
39. Ghosh, S. The New Wave: Renewable Energy and Global Energy Economics. 2015. Available online: https://www.linkedin.com/pulse/new-wave-renewable-energy-global-economics-sam-ghosh (accessed on 30 December 2017).

© 2019 by the authors. Licensee MDPI, Basel, Switzerland. This article is an open access article distributed under the terms and conditions of the Creative Commons Attribution (CC BY) license (http://creativecommons.org/licenses/by/4.0/).

Article

A Study of a Standalone Renewable Energy System of the Chinese Zhongshan Station in Antarctica

Yinke Dou [1],*, Guangyu Zuo [1], Xiaomin Chang [2] and Yan Chen [1]

[1] College of Electrical and Power Engineering, Taiyuan University of Technology, Taiyuan 030024, China; zuoguangyu0030@link.tyut.edu.cn (G.Z.); chenyanlxq@163.com (Y.C.)
[2] College of Water Resources Science and Engineering, Taiyuan University of Technology, Taiyuan 030024, China; changxiaomin@tyut.edu.cn
* Correspondence: douyinke@tyut.edu.cn; Tel.: +86-139-3464-6229

Received: 10 April 2019; Accepted: 7 May 2019; Published: 14 May 2019

Abstract: China has built four stations in Antarctica so far, and Zhongshan Station is the largest station among them. Continuous power supply for manned stations mainly relies on fuel. With the gradual increase in energy demand at the station and cost of fuel traffic from China to Zhongshan station in Antarctica, reducing fuel consumption and increasing green energy utilization are urgent problems. This research considers a standalone renewable energy system. The polar environments and renewable energy distribution of area of Zhongshan station are analyzed. The physical model, operation principle, and mathematical modeling of the proposed power system were designed. Low-temperature performance and state of charge (SOC) estimation method of the lead–acid battery were comprehensively tested and evaluated. A temperature control strategy was adopted to prevent the battery from low-temperature loss of the battery capacity. Energy management strategy of the power system was proposed by designing maximum power point tracking (MPPT) control strategies for wind turbine and PV array. The whole power system is broadly composed of a power generator (wind turbine and PV array), an uploading circuit, a three-phase rectifier bridge, an interleaved Buck circuit, a DC/DC conversion circuit, a switch circuit, a power supply circuit, an amplifier, a driver circuit, a voltage and current monitoring, a load, battery units and a control system. A case study in Antarctica was applied and can examine the technical feasibility of the proposed system. The results of the case study reveal that the scheme of standalone renewable energy system can satisfy the power demands of Zhongshan Station in normal operation.

Keywords: renewable energy; low-temperature energy storage; SOC; simulation

1. Introduction

The rapid changes of sea ice condition in Arctic and Antarctica in recent decades have been considered one of the most impactful phenomena on Earth [1–3]. There are more and more researches and observations organized by China in Antarctica every year. Up to now, China has four Antarctic research stations, namely the Great Wall Station (62°12′59″ S, 58°57′52″ W), the Zho2033ngshan Station (69°22′24″ S, 76°22′40″ E), the Kunlun Station (80°25′01″ S, 77°06′58″ E) and the Taishan Station (73°51″ S, 76°58′ E). The Great Wall Station and Zhongshan Station are both perennial research stations, which have the ability to accommodate dozens of expedition members and researchers to spend the whole year in Antarctica. Zhongshan Station was established in an area of Larsemann Hills on East Antarctica on 26 February 1989 as a Chinese observation base of high altitude physics, glaciers, atmosphere, ocean, biological ecology, geology, geomagnetism etc. Zhongshan Station is also the base camp for the Chinese National Antarctica Inland Research Base. Zhongshan Station can accommodate about 25 wintering personnel and 600 summer personnel. The current energy supply of Zhongshan Station mainly depends on fuel. The ecological environment in Antarctica is very fragile, and the fuel produces a lot of harmful gases. Although treated by Zhongshan Station, it cannot achieve zero pollution. The fossil fuels used by Zhongshan Station were transported by the Chinese observatory ship Xuelong every year. As the demand of research and observations increases in Antarctica, the amount of fuel consumption increases accordingly. Therefore, new demands for power supply in Antarctica for green, sustainable and less costly energy sources such as wind, solar, ocean were created.

When humans do research in Antarctica, the impact of human activities on the natural environment of the polar region should be minimized as much as possible. Based on the special meteorological condition of Antarctica, some demands need to be proposed during designing and building the power system used in Zhongshan Station, East Antarctica. The use of fossil fuels is limited due to the inevitable pollution. There are abundant resources of wind, solar, and ocean energy in Antarctica, which can be considered a great advantage in the development of environmentally friendly power generation. Some works and research have reported on small standalone hybrid wind–solar systems which are isolated from the grid for observing systems deployed in the field in the Arctic Ocean and Antarctica [4]. The design of the hybrid wind–solar system was adopted because of the fluctuations of the solar and wind energy resources in polar regions. Some researches of hybrid wind–solar systems have reported. A report by Reference [5] indicates that the best fit of the wind turbine and photovoltaic (PV) array to a given load can be determined by the least square method. Some methods of modeling, designing and evaluating of hybrid renewable energy systems were also developed [6]. A hybrid solar–wind–battery system was used in the isolated site of Potou in the northern coast of Senegal to realize the minimization of the annualized cost and loss of power supply probability [7]. Based on a methodology of optimal sizing of a hybrid PV/wind system, this hybrid power system was installed on Corsica and can meet the desired system reliability requirements [8]. In some harsh environments in Iraq, the design of hybrid systems can be considered renewable resources of power generation and the simulation results illustrate that it is possible to use the solar and wind energy to generate enough power for remote areas [9]. Renewable energy such as wind energy and solar energy have been used in Antarctica. Mawson Station of Australia has built two 300 KW wind turbines to provide continuous power since 2003. On Ross Island in Antarctica, a wind farm has been used to realize 100% of the energy supply of Scott Base of New Zealand and part of the power requirements of McMurdo Station of United States of America. For Princess Elisabeth Station of Belgium, 300 m^2 solar panels have been installed and can generate 49 MWh [10]. Syowa Station of Japan has built 55 KW of solar panels to produce an annual output of 44,000 KW h for accommodating up to 110 people in the summer and 28 people in the winter [11].

For a hybrid PV/wind system in Polar Regions, an energy storage system (ESS) plays an important role in storing excess energy and releasing the power as a reliable back-up to the power system for unpredictability and weather dependence of wind and solar energy. An integrated wind–PV hybrid system with a battery ESS was proposed and a power management strategy of this system can realize rapid control of the outputs of wind and PV power for regulating the battery current [12]. In the design of an isolated renewable hybrid power system, a methodology of battery sizing was used to determine the sizing curve and the feasible design space [13]. Using solar, wind, fuel cell, and batteries as input sources may be able to meet the load demand and an energy management strategy is proposed for a DC microgrid [14,15]. A case study of a stand-alone photovoltaic (PV) system was proposed and the environmental impact of batteries used in the renewable energy system was evaluated [16]. Due to the different seasonal changes of power production and demand, the design of renewable energy system should involve the use of surplus energy [17]. A multi-energy system with seasonal storage was designed and optimized in terms of total annual costs and carbon dioxide emissions [18].

In this study, a new standalone renewable energy system of the Chinese Zhongshan Station in Antarctica was designed to realize an environmentally friendly energy supply and to obtain high power generation efficiency. The physical model and mathematical model of the standalone renewable energy system were proposed [19]. Lead–acid batteries were selected as an energy storage system for the standalone hybrid windsolar system and a temperature control strategy was adopted to prevent the battery from low-temperature loss of the battery capacity. Energy management strategy of the power system was also proposed based on results of low-temperature characteristics of battery. The whole power system is broadly composed of 13 parts: (1) a power generator (wind turbines and PV array), (2) an uploading circuit, (3) a three-phase rectifier bridge, (4) an interleaved Buck circuit, (5) a DC/DC conversion circuit, (6) a switch circuit, (7) a power supply circuit, (8) an amplifier, (9) a driver circuit, (10) a voltage and current monitoring, (11) a load, (12) battery units, (13) a control system. A case study of the operational results of the standalone renewable energy system was examined to evaluate the technical feasibility and stability. Analysis of simulation operation results, emission reduction and costs and benefits of renewable energy applications in Antarctica were completed.

In these contexts, this paper focuses on exploring a standalone renewable energy system for Zhongshan Station. Section 2 describes the atmospheric conditions of the study area. Section 3 gives the results of the physical model, operation principle, and mathematical modeling of the power system. The results of the low-temperature characteristics of batteries are shown in Section 4. The energy management strategy of hybrid wind–solar system is given in Section 5. Section 6 introduces the design of the whole power system. The results of a case study of the hybrid wind–solar power system in Zhongshan Station are presented in Section 7. The conclusions are in final section.

2. Atmospheric Conditions of Zhongshan Station, East Antarctica

The meteorological data of Zhongshan Station were obtained from a manned weather station in 2015. Table 1 summarizes the various sensors used in the manned weather station. The weather station consists of a wind speed and wind direction detection sensor (Wind Monitor Model 05103-45, R.M.Young, Traverse City, MI, USA), a temperature and humidity sensor (HMP155A, Vaisala, Vantaa, Finland), an atmospheric pressure sensor (PTB110, Vaisala, Vantaa, Finland). The Wind Monitor sensor has a rugged and corrosion-resistant construction which is suitable for wind measuring applications in harsh environments. The four blade helicoid propeller of the wind speed sensor produces an AC sine wave voltage by rotation. The vane angle of the wind direction sensor is sensed by a precision potentiometer. The Wind Monitor sensor mounts on standard one-inch pipe. The temperature and humidity sensor has excellent stability and can withstand harsh environments. The temperature and humidity probe is protected with a sintered Teflon filter and a radiation shield, which can increase its lifetime by waterproofing, sandproofing and dustproofing. The atmospheric pressure sensor (PTB110) with the capacitive detection principle is a silicon capacitive absolute pressure sensor, which combines the outstanding elasticity characteristics and mechanical stability of single-crystal silicon.

Table 1. Sensor information of the manned weather station.

Sensor Name	Performance	Sensor Model
Wind speed	Temperature range: −60–30 °C Accuracy: 0.5 m/s	Wind Monitor Model 05103-45, R.M.Young, Traverse City, MI, USA
Wind direction	Temperature range: −60–30 °C Accuracy: 0.3°	Wind Monitor Model 05103-45, R.M.Young, Traverse City, MI, USA
Air temperature	Temperature range: −60–10 °C Accuracy: 0.1 °C	HMP155A, Vaisala, Vantaa, Finland
Air humidity	Temperature range: −60–10 °C Accuracy: 2%RH	HMP155A, Vaisala, Vantaa, Finland
Atmospheric pressure	Temperature range: −60–30 °C Accuracy: 0.6 hPa	PTB110, Vaisala, Vantaa, Finland

The data of wind speed, wind direction, air temperature, relative humidity and air pressure measured by the weather station in Zhongshan Station from 1 December 2014 to 1 November 2015 are shown in Figure 1. The height of the wind speed and direction with respect to the ground is 10 m. During this period the mean wind speed was 9.8 m/s. The maximum and minimum values of wind speed were 35.9 m/s and 0 m/s, respectively (Figure 1a). The wind near Zhongshan Station was dominated by strong easterly winds, but there were westerly winds with lower wind speeds in February and March (Figure 1b). Surface winds at Zhongshan Station are generally consistent with the observations before. Air temperature of Zhongshan Station is shown in Figure 1c. The average air temperature was −11.18 °C. The maximum and minimum values of air temperature were 8.30 °C and −39.9 °C, respectively. The lowest air temperature occurred on 8 July 2015 and the highest was on 20 December 2014. The results of air temperature distribution are similar to the previous observations. The average relative humidity was 59.6%. The maximum and minimum values of relative humidity were 96% and 26%, respectively. The highest relative humidity occurred on 25 October 2015 and the lowest was on 26 November 2015. As shown in Figure 1d, Zhongshan Station has low relative humidity and dry air. The average atmospheric pressure was 982.4 hPa. The maximum and minimum values of atmospheric pressure were 1013.3 hPa and 942.3 hPa, respectively. The lowest atmospheric pressure occurred on 11 April 2015 and the highest was on 18 July 2015 (Figure 1e). The short-term variations in wind speed, wind direction, air temperature, relative humidity and atmospheric pressure were considerable, which can prove the complexity of the weather conditions at Zhongshan Station.

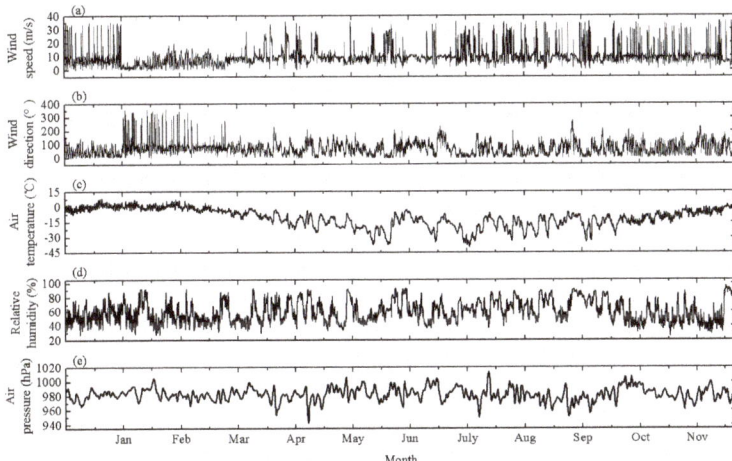

Figure 1. Time series of hourly (**a**) wind speed, (**b**) wind direction, (**c**) air temperature, (**d**) relative humidity and (**e**) air pressure obtained by the manned weather station at Zhongshan Station, Antarctica.

The multi-year average meteorological data are presented in Figure 2. The monthly average wind speed, radiation, day length, and air temperature in Figure 2 were obtained from the Atmospheric science data center of NASA. The monthly average wind speed from NASA and observed wind speed are shown in Figure 2a. The trend of wind speed from NASA is consistent with the observed results. For the multi-year average wind speed, the maximum and minimum wind speeds were 9.88 m/s in June and 7.5 m/s in January, respectively. The maximum and minimum monthly-observed wind speeds were 10.9 m/s in December and 4.8 m/s in February, respectively. The annual average wind speed from NASA was 9 m/s, which was similar to the annual observed result (9.78 m/s). The monthly average radiation and day length at Zhongshan Station are shown in Figure 2b. The radiation and day length decreased from January, and reached the minimum values at the same time (June). Then the radiation and day length continued to increase until December. The monthly mean radiation in May, June and July were 0.05 KW h/m^2/day, 0.00 KW h/m^2/day, 0.01 KW h/m^2/day, respectively. Additionally, the monthly mean day lengths were 4.05 h, 0 h and 1.23 h, respectively. This phenomenon can indicate that polar night occurs from late May to late July in this region. The monthly mean radiation in January, November, and December were 6.05 KW h/m^2/day, 4.97 KW h/m^2/day, 6.69 KW h/m^2/day, respectively.

The monthly mean day lengths in January, November, and December were 24 h, 20.8 h and 24 h, respectively, which polar day lasts from late November to early February of the following year at Zhongshan Station. As can be seen from Figure 2c, the monthly average air temperature at Zhongshan Station exhibited obvious seasonal characteristics. The average yearly air temperature was −19 °C. The maximum and minimum monthly mean air temperature was −7.13 °C in January and −27.3 °C in July.

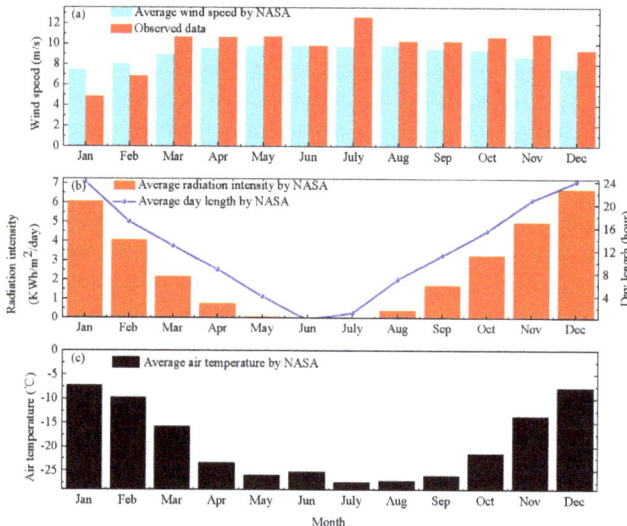

Figure 2. The monthly meteorological data of Zhongshan Station. (**a**) wind speed, (**b**) solar radiation and day length, (**c**) air temperature.

The weather condition of Zhongshan Station can be summarized as strong easterly winds, lower relative humidity, lower barometric pressure and cold air. The operating temperature range of the standalone renewable energy system in this study should be −50–30 °C based on the analysis of meteorological data at Zhongshan Station. Other meteorological elements should be taken into account during the design of the power supply system. Thus, in this study, the characteristics of power system at low temperatures should be considered and studied for achieve a long-term operation of research station in Antarctica.

3. System Design

3.1. Physical Model and Operation Principle

Based on the analysis of the atmospheric conditions, we designed the standalone renewable energy system. As shown in Figure 3, the proposed renewable energy system in this study is equipped with a power generator, an energy storage system, an end-user and a control station. The power generator consists of the PV arrays and wind turbines (WT), which can complete the conversion of wind energy and solar energy to electric energy. The energy storage system includes low-temperature batteries. The end-user is various loads at Zhongshan Station, which includes instruments for scientific research, electricity for daily use, heating, etc. The control station includes a control system, which has functions of controlling the process of charging and discharging of hybrid wind–solar power system. The rotation of wind turbines can produce AC currents. A three-phase rectifier circuit in the control system is designed to convert the three-phase alternating currents into stable direct currents. A DC chopper circuit is also designed to implement the control strategy for the renewable energy system to complete maximum power output. The control system is used to monitor voltage and current of

PV and WT, battery voltage and charging current. This proposed system would be a reliable and sustainable energy supply and guarantee the load demand of Zhongshan Station for 24 h a day.

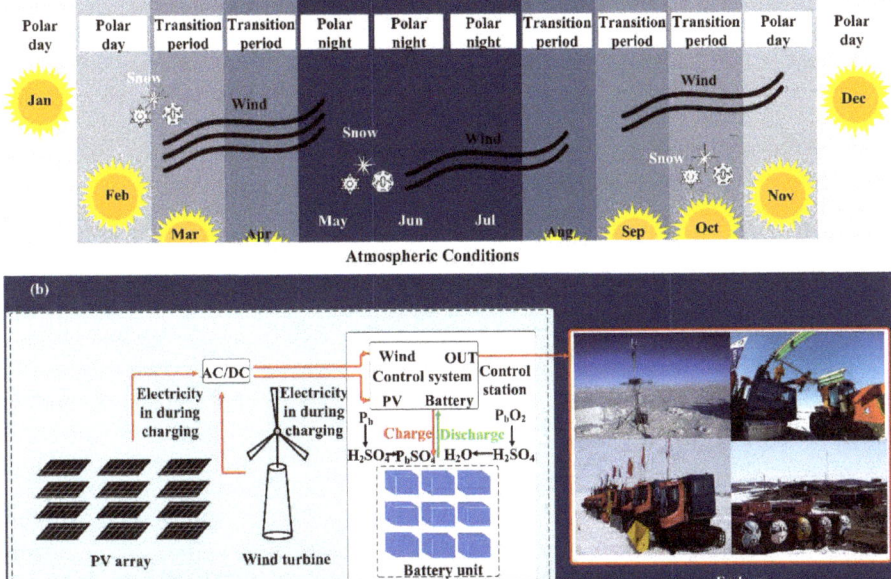

Figure 3. (**a**) Atmospheric conditions, (**b**) Schematic of the renewable energy system.

3.2. Mathematical Modeling of the Power System

3.2.1. PV Array

The polycrystalline panels were assembled by Taiyuan University of Technology in this study. The PV panels can be mainly divided into solar cells made of polymers, silicon materials, and sensitized nanomaterials [20] and silicon PV are mostly used. Advantages and disadvantages of different silicon solar cells are shown in Table 2.

Table 2. Advantages and disadvantages of different silicon solar cells.

Material Name	Advantages	Disadvantages
Monocrystalline silicon	High conversion efficiency, mature technology and small footprint	Expensive and high requirement for incident angle of sunlight
Polysilicon	The conversion efficiency is higher than that of amorphous silicon, the manufacturing cost is lower than that of monocrystalline silicon, and the low requirement for incident angle of sunlight	Conversion efficiency is lower than monocrystalline silicon, and the process is complicated
Amorphous silicon	Minimum requirement for incident angle of sunlight and high acceptance rate of astigmatism	Low conversion efficiency

As shown in Table 2, monocrystalline silicon solar cells and polycrystalline silicon solar cells have higher conversion efficiency and smaller size than amorphous silicon solar cells. The monocrystalline silicon solar cells and the polycrystalline silicon solar cells have different appearances due to different manufacturing processes. When assembled into PV panels, the monocrystalline silicon materials cannot be covered. In terms of efficiency of use, monocrystalline silicon solar cells and polycrystalline silicon solar cells are not much different, the former being 1%–2% more than the latter. Due to the different manufacturing processes used, the polycrystalline silicon solar cells are cheaper to produce than the monocrystalline silicon solar cells. Thus, the polycrystalline silicon solar cells are used in this study.

All the PV panels were designed to be positioned in a fixed direction, facing north. The key specifications of the PV panels are presented in Table 3.

Table 3. Key specifications of the photovoltaic (PV) panels.

Characteristics	Value	Unit
Open circuit voltage (V_{oc})	42.64	V
Optimum operating voltage (V_{mp})	34.96	V
Short circuit current (I_{sc})	9.48	A
Optimum operating current (I_{mp})	8.59	A
Maximum power at STC [1] (P_{max})	300	W
Operating temperature	−50 to 85	°C
Size	1956 × 992 × 50	mm

[1] Standard test conditions.

A total of 350 PV panels can be used to form a 120 V, 105 KW PV array. The principle of power generation of solar cell is that the solar radiation emits photons to the induction plate of the photovoltaic cell to produce a photoelectric effect, causing internal electrons to move, thereby generating current. Equivalent circuit of the solar cell is shown in Figure 4.

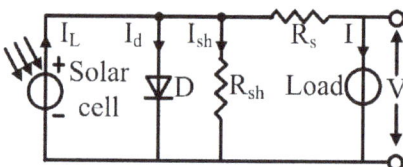

Figure 4. Equivalent circuit of the solar cell.

According to equivalent circuit of the solar cell, relevant calculating equations are as follows:

$$I = I_L - I_d - I_{sh} \qquad (1)$$

$$I_{sh} = \frac{IR_{sh} + V}{R_{sh}} \qquad (2)$$

The characteristics of the internal PN junction of the solar cell can be described as follows:

$$I_d = I_0 \left\{ \exp\left[\frac{q(IR_{sh} + V)}{\lambda KT}\right] - 1 \right\} \qquad (3)$$

Substituting Equations (2) and (3) into Equation (1) for calculation, relevant calculating equation is as follows:

$$I = I_L - I_0 \left\{ \exp\left[\frac{q(IR_{sh} + V)}{\lambda KT}\right] - 1 \right\} - \frac{IR_{sh} + V}{R_{sh}} \qquad (4)$$

where I is the output current (A); I_L is the photogenerated current (A); I_0 is the diode saturation current (A); q is the unit charge (1.6022×10^{-19} C); R_{sh} is the series resistance (Ω); V is the output voltage (V); λ is the diode ideality factor; K is Boltzmann's constant (1.3806×10^{-23} J/K); T is the cell temperature (K).

In this study, we use the following equations to describe the relationship between output of solar power and radiation intensity [4]:

$$\begin{cases} P_{solar} = P_{max}[1 - 0.004(T - T_{stc})]\beta_i \\ \beta_i = i\beta_1\beta_2\beta_3 \end{cases} \quad (5)$$

where P_{solar} is the output of solar power (Wh/day); P_{max} is the maximum power at standard test conditions (300 W); T is the ambient temperature (°C); T_{stc} is the ambient temperature at standard test conditions (25 °C); β_i is the adjustment parameter, which i is the average radiation intensity (KW h/m²/day), β_1 is the soiling losses factor 0.97, β_2 is the non-MPPT point coefficient 0.96, β_3 is the anti reverse diode coefficient 0.98.

3.2.2. Wind Turbine

The wind turbine designed and assembled by Taiyuan University of Technology was employed in this study. The key specifications of the wind turbine are presented in Table 4.

Table 4. Key specifications of the wind turbine.

Characteristics	Value	Unit
Cut in speed	2.5	m/s
Rated wind speed	10	m/s
Cut off speed	45	m/s
Rated power	10	KW
Peak power	12	KW
Diameter of impeller	7.8	m
Operating temperature	−50 to 85	°C
Number of blades	3	
Generator type	Three-phase AC permanent magnet generator	
Blade material	Reinforced glass steel	

Ten wind turbines can be used to form a 100 KW wind farm. Different types of wind turbines output different power based on their power curve characteristics. Through a comprehensive literature review, a model used to describe the performance is proposed as follows [19].

$$\begin{cases} P_{wind} = P_1 + P_2 + P_3 \\ P_1 = \sum P_R t(v/v_R)^3 (v_c \leq v \leq v_R) \\ P_2 = P_R \sum t(v_R \leq v \leq v_F) \\ P_3 = 0 (v < v_c \text{ and } v > v_R) \end{cases} \quad (6)$$

where P_{wind} is the output of wind power (Wh/day), which consists of P_1, P_2 and P_3; V_c is the cut-in wind speed (2.5 m/s); V_R is the rated wind speed (10 m/s); V_F is the cut-off wind speed (45 m/s); V is the wind speed; P_R is the rated electrical power (10 KW), which is average energy at the wind speed of 10 m/s for one minute; t is the time (hours).

3.3. Zhongshan Station Load Data

The load of Zhongshan Station can be divided into: (1) The first type of load is the internal heating system. Once the heating system is not working properly, it will affect the normal life of all the staff of Zhongshan Station and the lives of all personnel will be threatened. Thus, such loads cannot be cut off. (2) The second type of load is electricity for scientific research equipment. There is a lot of scientific research equipment installed in Zhongshan Station to monitor the climate, biochemistry, crustal changes, and movement of Antarctica in real time and obtain valuable on-site observation data. A power outage of equipment may lead to the discontinuity of observations and the lack of integrity of data. Thus, we need to ensure the continuous supply of electricity of scientific research equipment. (3) The third type of load is electricity for daily use. Such loads include electricity for lighting, recreational activities, electronics, etc., where necessary, such loads may be considered for power outages.

4. Analysis of Energy Storage System

At present, most of the wind–solar hybrid power generation systems use secondary batteries that can be repeatedly charged and discharged as energy storage systems, and the electrical energy can be converted into chemical energy for storage. When using electrical energy, the stored chemical energy of batteries can be turned into electrical energy. When we choose a suitable energy storage device, the capacity of the energy storage device and its charge and discharge performance are mainly considered. The battery with high conversion efficiency and low loss is suitable for the design of the standalone renewable energy system. In addition, the maintenance cost and life of the battery should be also key factors in the design. Commonly used batteries include lead–acid batteries, nickel–hydrogen batteries, nickel–cadmium batteries, lithium–ion batteries and sodium–sulfur batteries [20]. The key performance comparisons of each battery are presented in Table 5.

Table 5. Key performance comparisons of each battery.

Classification	Electrolyte	Principle	Operating Temperature (°C)
Lead–acid batteries	Dilute sulfuric acid	Oxidation-reduction reaction	−50 to 70
Lithium–ion batteries	Organic lithium salt electrolyte	Ion migration	−50 to 70
Nickel–cadmium batteries	Potassium hydroxide aqueous solution	Oxidation-reduction reaction	−20 to 45
Nickel–hydrogen batteries	Potassium hydroxide aqueous solution	Oxidation-reduction reaction	−20 to 60
Sodium–sulfur batteries	Na-β-Al_2O_3	Chemical reaction	300 to 350

It can be seen from Table 5 that the operating temperature ranges of lead–acid battery and lithium–ion battery are more suitable than other batteries for the requirements of this system. Compared with lead–acid batteries, lithium–ion batteries have certain safety hazards and lithium–ion batteries cost more than lead–acid batteries. Thus, the lead–acid battery is selected to design energy storage system.

A lead–acid battery consists of an electrolyte, positive and negative electrodes. P_b is used as the negative active material of lead–acid batteries, P_bO_2 can be the positive active material, and the electrolyte is diluted H_2SO_4. The energy conversion principle of lead-acid batteries can be expressed by the following chemical reaction equations.

$$P_bO_2 + H_2SO_4 + P_b \xrightarrow{Discharge} P_bSO_4 + 2H_2O \tag{7}$$

$$P_bSO_4 + 2H_2O \xrightarrow{Charge} P_bO_2 + H_2SO_4 + P_b \tag{8}$$

Equations (7) and (8) can describe the discharge and charging process of the lead–acid battery, respectively. In the process of discharge, P_b on the negative electrode is oxidized to become P_bSO_4, and P_bO_2 on the positive electrode is reduced to form P_bSO_4. Diluted H_2SO_4 in the surrounding area as an electrolyte participates in chemical reactions, forming P_bSO_4 while producing H_2O.

In the process of charge, P_bSO_4 on the negative electrode is reduced to form P_b and P_bSO_4 on the positive electrode is oxidized to become P_bO_2. The concentration of H_2SO_4 in the surrounding area gradually recovers. The charging process of the lead–acid battery is not finished until P_bSO_4 of the positive and negative electrodes is completely reduced to P_b and P_bO_2.

In the process of charge and discharge of lead–acid batteries, the terminal voltage can be expressed as the following equations.

$$U = E + \Delta\varphi_+ + \Delta\varphi_- + IR \tag{9}$$

$$U = E - \Delta\varphi_+ - \Delta\varphi_- - IR \tag{10}$$

Equations (9) and (10) represent the change in terminal voltage during the charging and discharging processes, respectively. U is the terminal voltage of the lead–acid battery (V); E is the electromotive force of batteries; $\Delta\phi_+$ is the overpotential of positive electrode (V); $\Delta\phi_-$ is the overpotential of negative electrode (V); I is the charge or discharge current (A); R is the internal resistance of the battery (Ω).

The life of the battery directly determines the time when the power supply system runs stably. This lead–acid batteries used in this power system were developed by Taiyuan University of Technology. Based on the principles of battery array combination, the battery of 2 V, 3 KAh is extended and the battery pack of 60 V, 45 KAh is used as the energy storage device of the power system.

4.1. Study on Low-Temperature Characteristics of Battery

The activity of the electrolyte of the lead–acid battery is easily affected by low temperatures, resulting in a decrease in battery capacity. The power supply system operating in Antarctica requires a long-term constant temperature treatment of energy storage system. Reasonable storage temperature needs to be determined, so as to reduce the energy consumption caused by maintaining the constant temperature as much as possible. Thus, a study on battery characteristics at low temperatures was designed and implemented.

In order to study the low-temperature characteristics of lead–acid batteries, a battery capacity calibration experiment was designed. A low-temperature test chamber (MDF-86V340E, Zhongkeduling, Hefei, Anhui, China) was used to provide stable low-temperature environments from −50 °C to 0 °C. The key specifications of the low-temperature test chamber are presented in Table 6.

Table 6. Key specifications of the low-temperature test chamber.

Characteristics	Value	Unit
Rated power	668	W
Rated voltage	220	V
Temperature range	−60 to 86	°C
Temperature adjustment accuracy	0.1	°C
Effective volume	340	L
Noise level	52	dB

The ideal capacity of the battery to be tested is 2 V, 3 KAh under a normal temperature environment. The battery was discharged at a constant current of 15 A at different ambient temperatures from −50 °C to 0 °C. In addition, the battery capacity at different ambient temperatures can be obtained. The battery voltages were measured by an oscilloscope (MSO70404C, Tektronix, Beaverton, OR, USA). The discharge cut-off voltage was set as 1.6 V. The interval for the experimental temperature change was set to 10 °C. We obtained the correlation between battery capacity and voltage at −50 °C, −40 °C, −30 °C, −20 °C, −10 °C and 0 °C. At each ambient temperature, the battery continued to discharge at a constant current until the cutoff voltage was reached. During the experiment, the low-temperature test chamber could maintain the temperature. We took the average values of the voltage to minimize the statistical error and uncertainty. As the temperature decreased, the battery capacity also decreased. The standstill battery capacities were 98.42%, 98.11%, 97.62%, 96.77%, 95.83% and 94.12% at 0 °C, −10 °C,

−20 °C, −30 °C, −40 °C, −50 °C, respectively. The discharge capacity of the battery was weakening due to low temperatures. Therefore, the storage temperature of the battery needs to be kept above 0 °C to avoid the low-temperature loss of the battery capacity.

4.2. SOC Estimation

Usually, the battery state of charge and the remaining useful life are considered as two important parameters to quantify and monitor the present battery state. In this study, long-term low temperature is the main factor affecting the battery capacity and the remaining useful life. References [21,22] have reported joint/dual extended Kalman filter and unscented Kalman filter with an enhanced self-correcting model, which can simultaneously estimate the SOC and capacity. The SOC estimation in this study is realized based on study on low-temperature characteristics of battery. The relationship between battery voltage and battery capacity at low temperatures is shown in Figure 5.

The mathematical model of the relationship between battery voltage and battery capacity at different temperatures can be expressed as follows.

$$\begin{cases} U_{-50\,°C} = 5.5 \times 10^{-8} \times SOC_{-50\,°C}^6 - 2.51 \times 10^{-5} \times SOC_{-50\,°C}^5 + 4.72 \times 10^{-3} \times SOC_{-50\,°C}^4 - 0.471 \times SOC_{-50\,°C}^3 \\ + 26.24 \times SOC_{-50\,°C}^2 - 711.8 \times SOC_{-50\,°C} + 9362.2 \\ U_{-40\,°C} = 2.9 \times 10^{-8} \times SOC_{-40\,°C}^6 - 1.36 \times 10^{-5} \times SOC_{-40\,°C}^5 + 2.58 \times 10^{-3} \times SOC_{-40\,°C}^4 - 0.261 \times SOC_{-40\,°C}^3 \\ + 14.64 \times SOC_{-40\,°C}^2 - 434.1 \times SOC_{-40\,°C} + 5306.9 \\ U_{-30\,°C} = 1.3 \times 10^{-8} \times SOC_{-30\,°C}^6 - 6.04 \times 10^{-6} \times SOC_{-30\,°C}^5 + 1.15 \times 10^{-3} \times SOC_{-30\,°C}^4 - 0.116 \times SOC_{-30\,°C}^3 \\ + 6.487 \times SOC_{-30\,°C}^2 - 192.4 \times SOC_{-30\,°C} + 2353.8 \\ U_{-20\,°C} = 1.5 \times 10^{-8} \times SOC_{-20\,°C}^6 - 6.82 \times 10^{-6} \times SOC_{-20\,°C}^5 + 1.31 \times 10^{-3} \times SOC_{-20\,°C}^4 - 0.132 \times SOC_{-20\,°C}^3 \\ + 7.508 \times SOC_{-20\,°C}^2 - 244.2 \times SOC_{-20\,°C} + 2760.1 \\ U_{-10\,°C} = 8.0 \times 10^{-9} \times SOC_{-10\,°C}^6 - 3.72 \times 10^{-6} \times SOC_{-10\,°C}^5 + 7.13 \times 10^{-4} \times SOC_{-10\,°C}^4 - 0.072 \times SOC_{-10\,°C}^3 \\ + 4.098 \times SOC_{-10\,°C}^2 - 122.4 \times SOC_{-10\,°C} + 1508.8 \\ U_{0\,°C} = 4.66 \times 10^{-9} \times SOC_{0\,°C}^6 - 2.13 \times 10^{-6} \times SOC_{0\,°C}^5 + 4.01 \times 10^{-4} \times SOC_{0\,°C}^4 - 0.0401 \times SOC_{0\,°C}^3 + \\ 2.229 \times SOC_{0\,°C}^2 - 25.5895 \times SOC_{0\,°C} + 797.2 \end{cases} \quad (11)$$

where $U_{-50°C}$, $U_{-40°C}$, $U_{-30°C}$, $U_{-20°C}$, $U_{-10°C}$ and $U_{0°C}$ are the battery voltages at −50 °C, −40 °C, −30 °C, −20 °C, −10 °C, 0 °C, respectively; $SOC_{-50°C}$, $SOC_{-40°C}$, $SOC_{-30°C}$, $SOC_{-20°C}$, $SOC_{-10°C}$ and $SOC_{0°C}$ are the values of state of charge at −50 °C, −40 °C, −30 °C, −20 °C, −10 °C, 0 °C, respectively.

We evaluated the relationship between battery voltage and battery capacity from −50 °C to 0 °C. However, the values at the non-measured battery voltage and battery capacity could be predicted by interpolation. Equation (11) can be described as follows.

$$\begin{aligned} U(T_i) &= a(T_i)SOC^6(T_i) + b(T_i)SOC^5(T_i) + c(T_i)SOC^4(T_i) \\ &+ d(T_i)SOC^3(T_i) + e(T_i)SOC^2(T_i) + f(T_i)SOC(T_i) + g(T_i) \end{aligned} \quad (12)$$

where $U(T_i)$ is the battery voltages at different temperatures from −50 °C to 0 °C; $SOC(T_i)$ is the battery capacity at different temperatures from −50 °C to 0 °C; $a(T_i), b(T_i), c(T_i), d(T_i), e(T_i), f(T_i)$, and $g(T_i)$ are the coefficients of the Equation (12), which have dependences of low temperatures.

The coefficients $a(T_i), b(T_i), c(T_i), d(T_i), e(T_i), f(T_i)$ and $g(T_i)$ from −50 °C to 0 °C are shown in Figure 6.

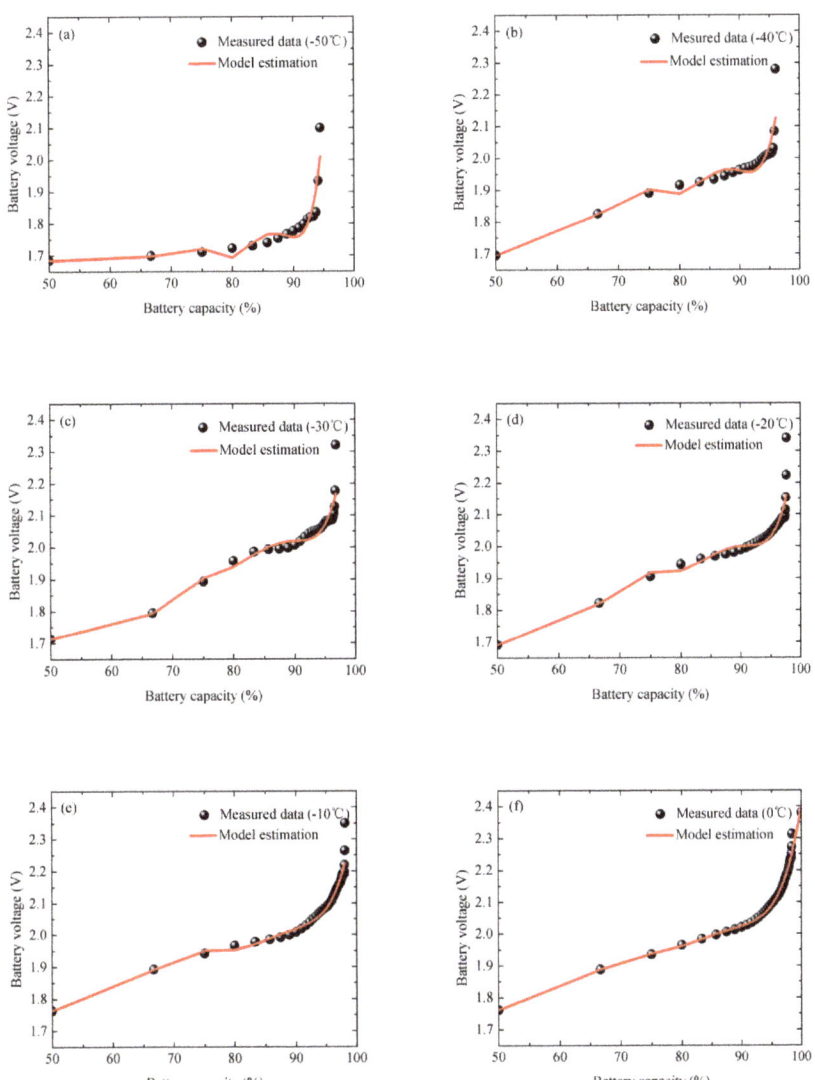

Figure 5. (**a**–**f**) show the relationship between battery voltage and battery capacity at −50 °C, −40 °C, −30 °C, −20 °C, −10 °C, 0 °C, respectively.

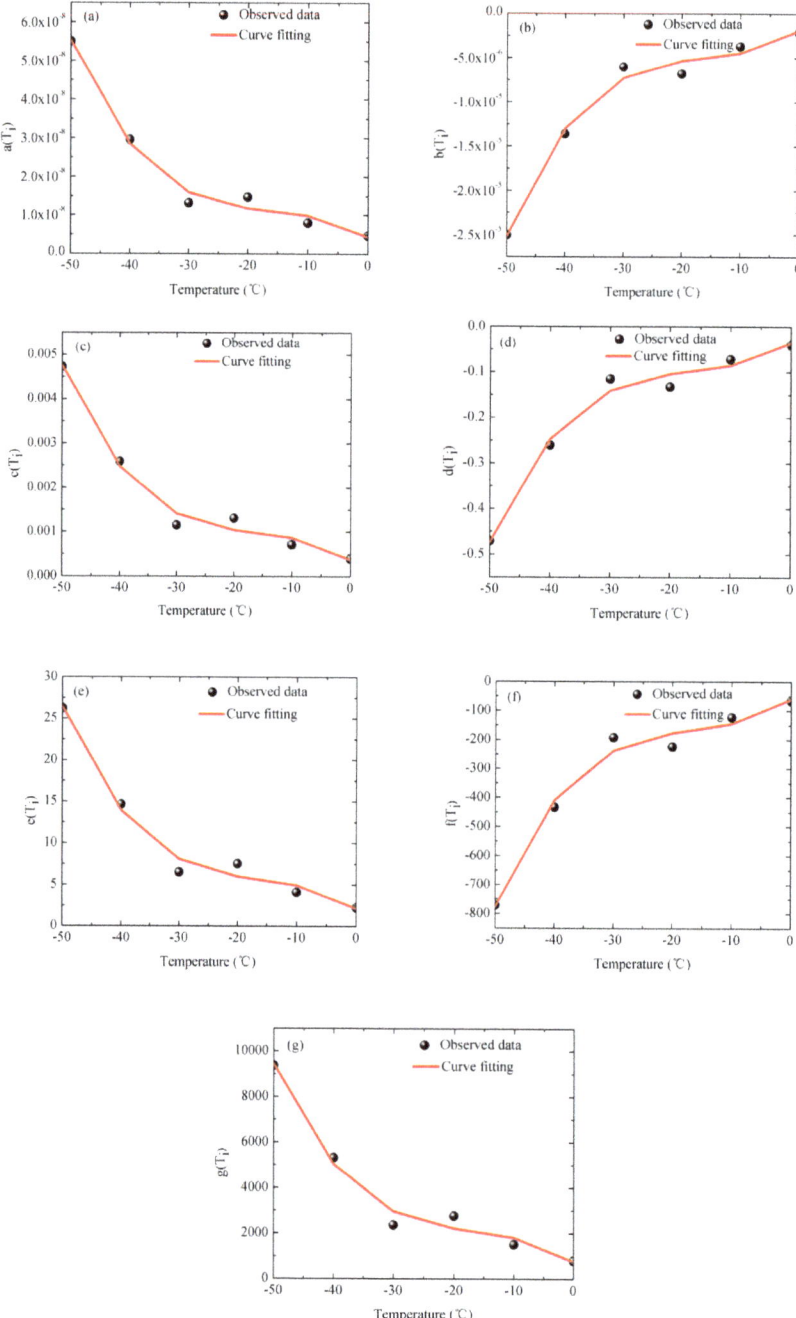

Figure 6. (a–g) show the coefficients $a(T_i)$, $b(T_i)$, $c(T_i)$, $d(T_i)$, $e(T_i)$, $f(T_i)$ and $g(T_i)$ *from* $-50\ °C$ to $0\ °C$, respectively.

As shown in Figure 6, the temperature dependence of the coefficients of $a(T_i)$, $b(T_i)$, $c(T_i)$, $d(T_i)$, $e(T_i)$, $f(T_i)$ and $g(T_i)$ is generally in the form of a cubic functions. The coefficients $a(T_i)$, $b(T_i)$, $c(T_i)$, $d(T_i)$, $e(T_i)$, $f(T_i)$ and $g(T_i)$ in Equation (12) at different temperatures could be predicted as follows.

$$\begin{cases} a(T_i) = -9.96 \times 10^{-13} T_i^3 - 4.8 \times 10^{-11} T_i^2 - 9.28 \times 10^{-10} T_i + 4.27 \times 10^{-9} \\ b(T_i) = 4.52 \times 10^{-10} T_i^3 + 2.19 \times 10^{-8} T_i^2 + 4.28 \times 10^{-7} T_i - 1.96 \times 10^{-6} \\ c(T_i) = -8.46 \times 10^{-8} T_i^3 - 4.11 \times 10^{-6} T_i^2 - 8.17 \times 10^{-5} T_i + 3.69 \times 10^{-4} \\ d(T_i) = 8.41 \times 10^{-6} T_i^3 + 4.1 \times 10^{-4} T_i^2 + 8.25 \times 10^{-3} T_i - 0.037 \\ e(T_i) = -4.66 \times 10^{-4} T_i^3 - 0.023 T_i^2 - 0.4657 T_i + 2.066 \\ f(T_i) = 0.0137 T_i^3 + 0.675 T_i^2 + 13.8716 T_i - 60.908 \\ g(T_i) = -0.1654 T_i^3 - 8.2046 T_i^2 - 170.3108 T_i + 741.803 \end{cases} \quad (13)$$

4.3. Temperature Control Strategy

Based on the analysis on Sections 4.1 and 4.2, the storage temperature of the battery needs to be kept above 0 °C to prevent the battery from low-temperature loss of the battery capacity. The heating system in the battery storage compartment ensures that the temperature of the battery can be constant within the ideal operating temperature range and reduces energy consumption by means of intermittent starting. The proportion-integral-differential (PID) algorithm was chosen to design the temperature control strategy.

The chosen heating system is a complex system with larger time lag and inertia. The mathematical model of the temperature control system in this study is described by a first-order inertia lag link. The transfer function of the heater can be expressed as follows.

$$G(s) = ke^{-\tau s}/(Ts+1) \quad (14)$$

where k is static gain; T is time constant; τ is pure lag time.

This study uses an incremental PID control algorithm. A step input signal is applied to the controlled object to measure the step response of the controlled object, and the approximate transfer function of the controlled object can be obtained by the flying up curve method. Parameters of the transfer function can be seen in Table 7.

Table 7. Parameters of the transfer function.

Parameter	Value
k	0.8
T	48.75
τ	11.2

The flowchart of the temperature control strategy is given in Figure 7.

After the PID was initialized, the target temperature for battery storage T_a was set first and then the real-time temperature of battery storage T_r was obtained. Generally, T_a is set to be higher than 0 °C. If $T_a < T_r$, the battery storage temperature could be considered suitable. If $T_a > T_r$, the heating system will be started and the difference between the target temperature for battery storage T_a and the real-time temperature of battery storage T_r will calculated, which is marked as e. If $e < 3$, the incremental PID control algorithm will be used to heat the battery storage room. If $e > 3$, the full power heating will be activated to quickly reach the target temperature for battery storage. After heating, the difference between T_a and T_r will be evaluated again. If $T_a \leq T_r$, the heating system will end the heating of the battery storage compartment.

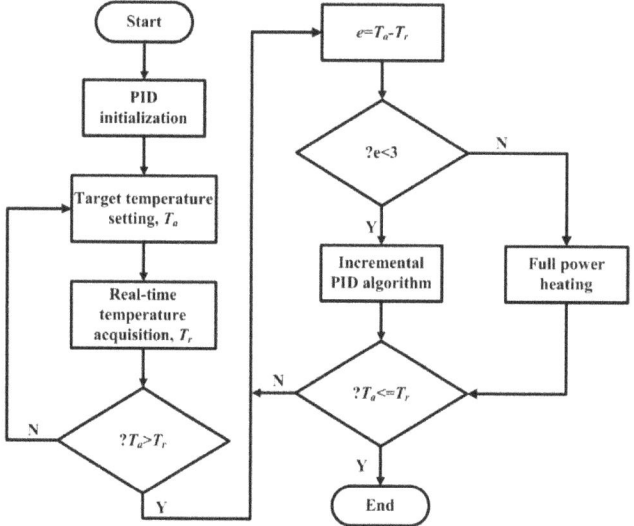

Figure 7. The flowchart of the temperature control strategy.

5. Energy Management Strategy

5.1. MPPT Control Strategy for Wind Turbine

Perturbation and observation method with variable step is used to achieve maximum power output of the wind turbine by adjusting the duty ratio. The method needs to estimate the position of the current maximum power point by real-time monitoring of the power difference between the two times, thereby determining the size of the duty cycle. If the difference is positive, the duty cycle will be decreased. Otherwise, the duty cycle will be increased. If it is zero, it indicates that the maximum power point has been reached. Since the wind speed in nature is randomly fluctuating, it may cause the power supply system to oscillate. Two step sizes are proposed in this study. The threshold of the power difference is set. If the difference is within the threshold, a small step will be used. If the difference is outside the threshold, a large step will be used to gradually approach the maximum power point.

The mechanical energy produced by the wind turbine in this study can be expressed as follows.

$$P_{WT} = \frac{1}{2} C_p(\lambda, \beta) S \sigma v^3 \tag{15}$$

where P_{WT} is the output of wind turbine (W); $C_p(\lambda, \beta)$ is wind energy utilization factor; λ is tip speed ratio; β is pitch angle of blade; S is sweep area (m^2); σ is air density (Kg/m^3); v is wind speed (m/s).

The sweep area of wind turbine can be described as follows.

$$P_{WT} = \frac{1}{2} C_p(\lambda, \beta) S \sigma v^3 \tag{16}$$

where R is the impeller radius (m).

The wind energy utilization factor can be described as follows.

$$C_p(\lambda, \beta) = 0.5176(116\lambda_c - 0.4\beta - 5)e^{-\frac{21}{\lambda_c}} + 0.0068\lambda \tag{17}$$

λ and λ_c in Equation (16) can be described as follows.

$$\begin{cases} \lambda = \frac{\omega R}{v} = \frac{2\pi n R}{60 v} \\ \lambda_c = \frac{1}{\lambda + 0.08\beta} - \frac{0.035}{\beta^3 + 1} \end{cases} \quad (18)$$

where ω is wind turbine angular velocity (rad/s); n is rotational speed of wind turbine (r/min).

In the study of the MPPT strategy of the wind turbine, β (pitch angle of blade), v (wind speed) and ω (wind turbine angular velocity) are set as input to the wind turbine power generation model. Air density is selected as 0.927 Kg/m^3. β is 0 in this study. For wind speed v, we built a natural wind speed model. The wind speed model is considered as a combination of basic wind speed V_b, gradual wind speed V_r, and gust wind speed V_g. The basic wind speed V_b is the average wind speed (7 m/s). The gradual wind speed V_r characterizes the slow change of the wind speed and can be expressed by follows.

$$V_r = V_r(\max) \frac{t_{r1} - t}{t_{r1} - t_{r2}} \quad (19)$$

where $V_r(max)$ is maximum value of gradual wind speed (10 m/s); t_{r1} is start time of gradual wind (4 s); t_{r2} is end time of gradual wind (11 s); t is time of gradual wind.

The gust wind speed V_g can characterize the degree of abrupt change in wind speed and can be expressed by follows.

$$V_g = V_g(\max)/2 \times \left[1 - \cos 2\pi\left(\frac{t - t_{g1}}{T_g}\right)\right] \quad (20)$$

where $V_g(max)$ is maximum value of gust wind speed (6 m/s); t_{r1} is start time of gust wind (3 s); T_g is period of gust wind (6 s); t is time of gust wind.

The model of natural wind speed can be described as follows.

$$v = V_b + V_r + V_g \quad (21)$$

We introduced the natural wind model into the wind power model, and the simulation results obtained by MATLAB Simulink of the wind speed, wind turbine output power and rotational speed of wind turbine are shown in Figure 8.

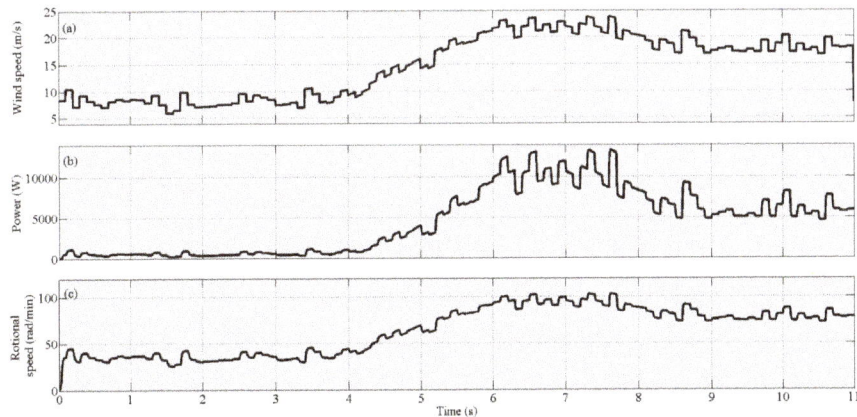

Figure 8. The simulation results of (**a**) the wind speed, (**b**) wind turbine output power and (**c**) rotational speed of wind turbine.

As can be seen in Figure 8a, at the beginning, the wind speed was low (about 2–5 m/s), thus the output power of the wind turbine was also low (Figure 8b). From 4 s to 6 s, the wind speed increased rapidly from the beginning of 2 m/s to 17 m/s, and the corresponding wind turbine output power was gradually increased, which the maximum power can reach 13 kW. After the 8 s, the wind speed began to slowly decrease and the output power decreased slowly. We also simulated the effects of abrupt wind speeds on wind turbine output power at 6 m/s, 8 m/s and 10 m/s (Figure 9). Under the influence of abrupt wind, the response time of wind turbine speed and power is less than 0.14 s.

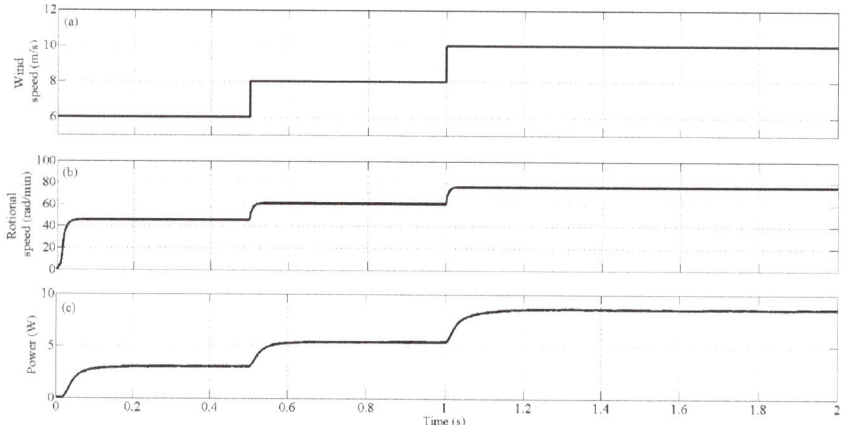

Figure 9. The simulation results of (a) the wind speed, (b) wind turbine output power and (c) rotational speed of wind turbine at abrupt wind speeds.

5.2. MPPT Control Strategy for PV Array

Perturbation and observation method with adaptive variable step is adopted, which achieves self-selection of the step size by adding an adaptive algorithm when setting the step size. This method not only improves the steady state performance of the power system, but also improves the dynamic performance. The step size calculation can be described as follows.

$$S(k+1) = N \frac{P(k) - P(k-1)}{S(k)} \tag{22}$$

where $S(k)$ is step size ($0 < S(k) < 1$); N is the constant determined by the sensitivity of the adaptive variable step size adjustment; $P(k)$ is power.

The flowchart of the perturbation and observation method with adaptive variable step is given in Figure 10.

The perturbation and observation method with adaptive variable step obtained $I(k-1)$, $I(k)$, $U(k-1)$ and $U(k)$ to calculate $P(k-1)$ and $P(k)$. We got the value of difference of power dP (Figure 10). The threshold of the power difference (e_p) was set. The direction of perturbation can be determined by calculating $dP-e_p$. Then the step size $S(k+1)$ can be adjusted by Equation (21). Until $dP = 0$, the maximum power point can be considered to be reached.

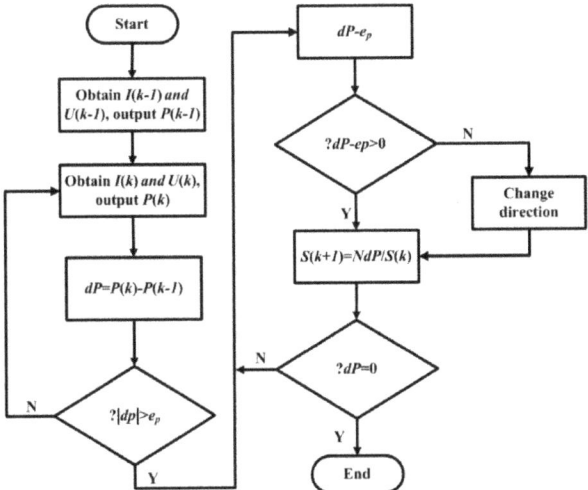

Figure 10. The flowchart of the perturbation and observation method with adaptive variable step.

5.3. Power Supply Strategy

The prerequisite for stable operation of the wind–solar hybrid power system is to maintain the energy balance between power generation and power consumption. If the power converted by wind and solar is less than the actual load, the battery plays the role to supply power to the load. On the contrary, if the power converted by wind and solar is greater than the actual load, the battery stores excess electrical energy. The flowchart of the power supply strategy is given in Figure 11.

After the energy assessment, three energy supply methods were selected: (1) Wind energy available; (2) Wind and solar energy available; (3) Solar energy available. After the SOC estimation was completed, whether the output power of the generator (wind turbine and PV array) can meet the load was calculated and evaluated. In all three energy supply methods, when the energy output is greater than the load, the output power of the generator can supply the load. If SOC < 90%, the output power of the generator should charge the battery. Otherwise, the battery does not need to be charged. When the energy output is less than the load, as long as SOC is greater than 10%, the battery and the power generator can directly provide the power of the load. If SOC < 10%, the power supply of a part of the load should be cut off to ensure the power generator and the battery providing power to the remaining load.

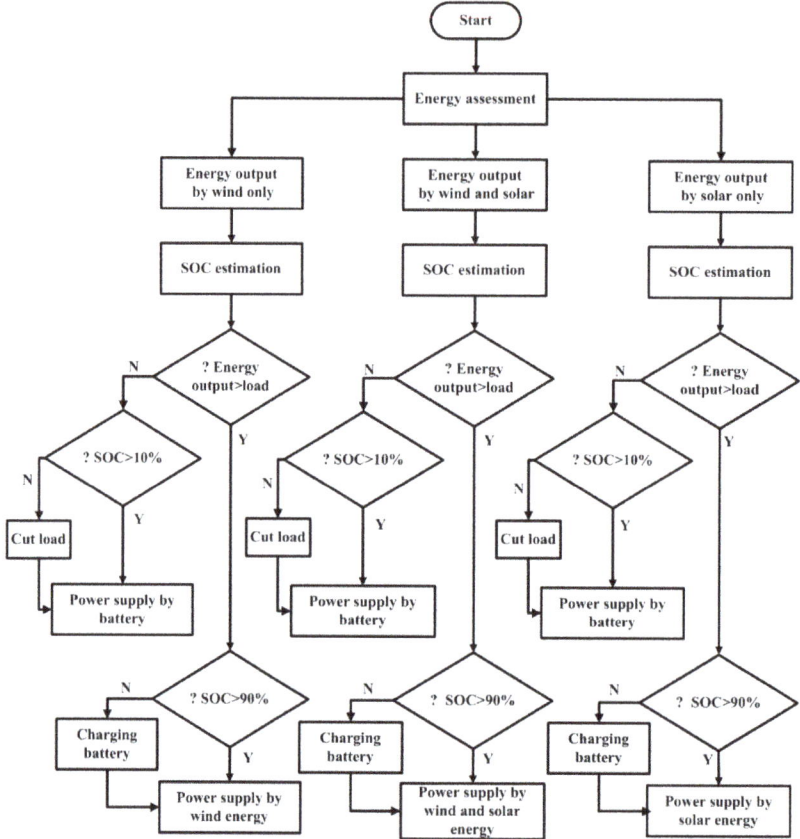

Figure 11. The flowchart of the power supply strategy.

6. Circuit Design

The block diagram of the standalone renewable energy system is illustrated in Figure 12. As can be easily seen, the whole system is broadly composed of a power generator (wind turbines and PV array), an uploading circuit, a three-phase rectifier bridge, an interleaved Buck circuit, a DC/DC conversion circuit, a switch circuit, a power supply circuit, an amplifier, a driver circuit, a voltage and current monitoring, a load, battery units, and a control system. The electric energy generated by the wind turbine converts into direct current through the three-phase rectifier bridge, and concentrates with the electric energy generated by the photovoltaic power generator. Then the electric energy generated by the wind energy and the solar energy are converted into stable direct currents to loads and the battery units by the interleaved Buck circuit. The wind turbine side is designed with an unloading circuit to prevent excessive output power from damaging the equipment under high wind conditions. The bus voltage is 24 V and the battery voltage is 12 V, thus, a DC/DC conversion circuit is designed. The model's parameters of circuits were chosen by empirical and commercial general specifications, which is a common method in circuit design.

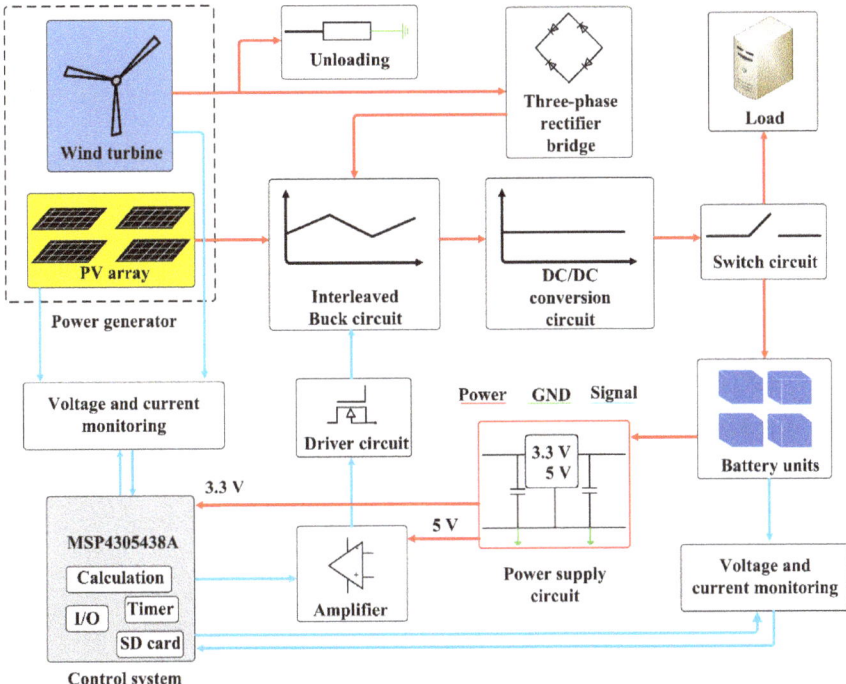

Figure 12. Block diagram of the standalone renewable energy system by hybrid solar–wind system.

6.1. Control System

In this study, the core of control system is selected as MSP4305438A (Texas Instruments, Dallas, TX, USA). The microcontroller is a reduced instruction set computer (RISC) with a 16-bit mixed-signal processor, which has a high processing power, a fast computing speed and an efficient development environment. The microcontroller can operate stably in a low-temperature environment (−50 °C) and was used multiple times in monitoring in Antarctica and the Arctic Ocean [4]. The control system has 11 sets of I/O ports, which can monitor the voltage and current of wind turbine, PV array and batteries in real time, and realizes the control strategy of the power system. Terminal names and general descriptions of the electrical interface of the control system are shown in Table 8.

Table 8. Terminal names and general descriptions of the electrical interface of the control system.

Terminal Name	Description	Direction
P6.7	Voltage monitoring	Input
P7.5	Current monitoring	Input
P6.0	Battery condition monitoring	Input
P6.1	Wind turbine unloading	Output
P1.1, P1.2, P4.5, P4.6	PWM control	Output
P4.0, P4.1, P4.2, P4.3	PWM drive	Output

6.2. DC/DC Conversion Circuit and Power Supply Circuit

The DC/DC conversion circuit and power supply circuit are shown in Figure 13. In this study, the DC/DC conversion circuit can provide better stability to the proposed power system over wide ranges of input and output voltages, and enable more stable and accurate current limiting operation. A thermal shutdown in this circuit is implemented to prevent damages owing to excessive heat.

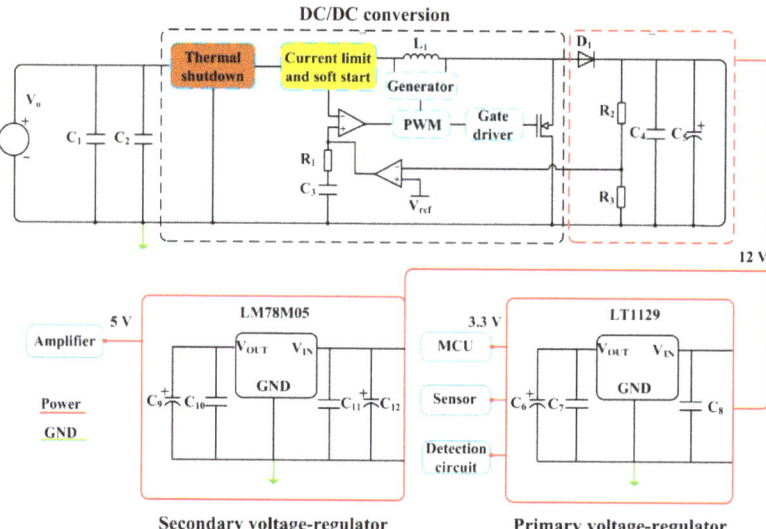

Figure 13. The DC/DC conversion circuit and power supply circuit.

The output voltage of DC/DC conversion circuit can be set by external resistors and the resistors are calculated as follows.

$$R_2 = V_{ref}/70\mu A \tag{23}$$

$$R_3 = R_2\left(12/V_{ref} - 1\right) \tag{24}$$

where R_2 and R_3 are external resistors; V_{ref} is the reference voltage set inside the circuit (1.238 V). The resistances of R_2 and R_3 are 18 KΩ and 156 KΩ, respectively. Other relevant parameters of the DC/DC conversion circuit can be seen in Table 9.

Table 9. Relevant parameters of the DC/DC conversion circuit.

Parameter	Value	Unit
C_1	1	µF
C_2	1	µF
C_3	3.3	nF
C_4	0.1	µF
C_5	10	µF
R_1	13	KΩ
R_2	18	KΩ
R_3	156	KΩ
L_1	3.3	µH

The power supply circuit includes a primary voltage-regulator and a secondary voltage-regulator. Primary voltage-regulator is designed by LT1129 (Analog Devices Inc., Norwood, MA, USA), which can generate 3.3 V at supply of 12 V for MCU, some sensors and some detection circuits in this power system. LM78M05 (Texas Instruments, Dallas, TX, USA) is selected as the core of the secondary voltage-regulator, which can generate 5 V at supply of 12 V to fulfill the requirements of the amplifier. Relevant parameters of the power supply circuit can be seen in Table 10.

Table 10. Relevant parameters of the power supply circuit.

Parameter	Value	Unit
C_6	100	μF
C_7	0.1	μF
C_8	0.1	μF
C_9	100	μF
C_{10}	0.1	μF
C_{11}	0.1	μF
C_{12}	0.47	μF

6.3. Charging Circuit

The solar/wind charging circuit is shown in Figure 14. The three-phase input of wind turbine converts AC to DC through a three-phase rectifier bridge, and mixes with the input of the PV array. In this circuit, freewheeling diodes are replaced to reduce rectification losses and the current sharing effect is significantly enhanced. The electric energy generated by wind and solar is very unstable. The electric energy is converted into a stable and usable DC through the interleaved Buck circuit. The choice of MOS tube considers the following elements: (1) Withstand voltage greater than or equal to 3 times DC bus voltage; (2) The current value is less than 1/4 of the rated current; (3) Low on resistance.

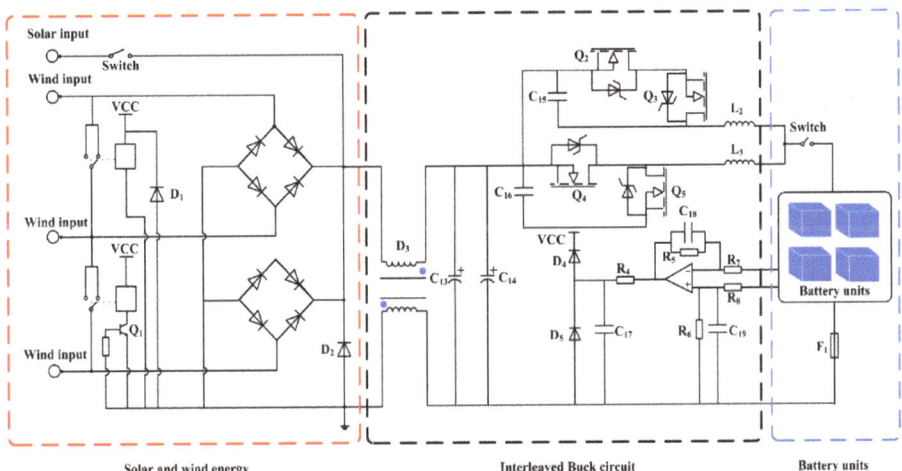

Figure 14. The solar/wind charging circuit.

As shown in Figure 14, MOSFET RU190N08 (Ruichips Semiconductor, Shenzhen, Guangdong, China) is selected as the power device of the interleaved Buck circuit, which has a withstand voltage of 80 V and a withstand current of 190 A. Relevant parameters of the power supply circuit can be seen in Table 11.

The inductance of interleaved Buck circuit can be obtained as follows.

$$L_{Buck} = V_0(1-D)/(rI_L f_s) \tag{25}$$

where L_{Buck} is the inductance of interleaved Buck circuit; V_0 is DC bus voltage (24 V); D is the minimum duty cycle of the PWM control method (0.57); r is inductor current ripple peak-to-peak factor (0.4); I_L is inductor rated current (40 A); f_s is switching frequency of PWM control signal (20 KHz). Thus the inductance of interleaved Buck circuit L_{Buck} is 32 μH.

Table 11. Relevant parameters of the solar/wind charging circuit.

Parameter	Value	Unit
C_{13}	470	μF
C_{14}	330	μF
C_{15}	100	pF
C_{16}	100	pF
C_{17}	0.1	μF
C_{18}	0.1	μF
C_{19}	0.1	μF
R_4	30	KΩ
R_5	15	KΩ
R_6	15	KΩ
R_7	30	KΩ
R_8	30	KΩ
L_2	100	mH
L_3	100	mH

7. A case Study in Antarctica

7.1. Existing Power Supply System in Zhongshan Station

At present, there are three sets of diesel generator in Zhongshan Station. In the process of power generation, the on-duty personnel should be arranged to record and maintain the relevant parameters of the generators. Based on the data provided by the Chinese National Antarctica Research Expeditions, we aggregated the data of the existing diesel power supply system at Zhongshan Station in Antarctica in 2013, 2014 and 2015. The average monthly power supply and monthly fuel consumption of Zhongshan Station are obtained. The actual monthly load power and fuel consumption are shown in Figure 15.

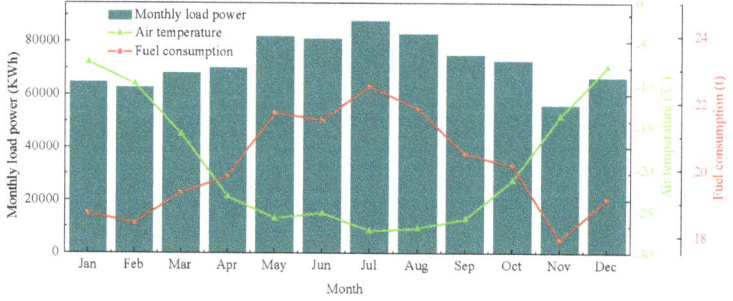

Figure 15. Actual monthly load power, air temperature and fuel consumption in Zhongshan Station.

As can be seen from Figure 15, the annual power consumption of Zhongshan Station is closely related to climatic conditions. The trend of monthly load power has a good correlation with the monthly average air temperature. The peak annual power consumption of Zhongshan Station is concentrated in May, June, July and August. At this time, it is the winter and polar night in Antarctica, which the air temperature in these months is the lowest in one year. However, scientists would continue to conduct scientific investigations near Zhongshan Station. At the same time, the outside temperature is low and the demand of indoor heating is increasing. Therefore, the power consumption of Zhongshan Station will also increase. As shown in Figure 2b, the annual radiation intensity and day lengths are the lowest in this period. The power supply of Zhongshan Station mainly depended on the wind turbine and the battery. Zhongshan Station has a total load of 100 KW. The monthly average power consumption is 72,516 KW h, the monthly average fuel consumption is above 20 t. Polar day occurs from December to January in Zhongshan Station, and it can make full use of solar power to generate electricity. The power consumption of the load in February, March, April, September, October, and November basically maintain the average power consumption of Zhongshan Station, which both wind and solar energy resources can generate electricity, and batteries have electricity reserves.

7.2. Analysis of Simulation Operation Results

The power generation of the standalone renewable energy system in Zhongshan Station from 2014 to 2015 is presented in Figure 16.

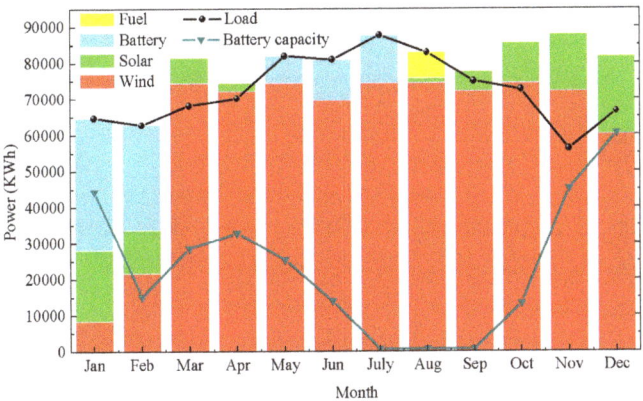

Figure 16. The power generation of the standalone renewable energy system in Zhongshan Station from 2014 to 2015.

The results of Zhongshan Station load were previous monitoring data. The power generated by wind energy and solar energy was calculated by mathematical models of the standalone renewable energy system. If the power generated by the standalone renewable energy system cannot meet the requirement of load, the battery will act as an energy storage system to supply the load of the Zhongshan Station to maintain the normal operation. Accordingly, battery capacity can be also derived from the simulation. The observed wind speed data were selected to complete simulation calculation. The values of radiation intensity from NASA were used to calculate power generation by PV array. As can be seen in Figure 16, the monthly average load of Zhongshan Station remained basically stable within one year. The annual load of Zhongshan Station was 870,196 KW h. The minimum and maximum values of the load were 56,175 KW h in November and 87,743 KW h in July, respectively. The average value of the load of Zhongshan Station was 72,516 KW h. Wind power production dominated the power supply of the standalone renewable energy system and generated 747,858.4 KW h in a year, which was 86% of the load. The monthly average maximum and minimum values of wind power were 74,400 KW h in March, May, July, August, October and 8455.4 KW h in January. In March, April, October and November, only the energy generated by wind can meet the load of Zhongshan Station. The solar power was shown strong seasonal fluctuations owing to the polar night (June and July) in Antarctica. The annual solar power of Zhongshan Station was 97,361 KW h. The monthly average solar power of polar day in January and December were larger, which were 19,659.9 KW h and 21,779.7 KW h, respectively. The minimum value of monthly average solar power in a year appeared in June (0 KW h) and July (34 KW h). In January, February, May, June and July, wind and solar energy were less than the load required for the month, thus, the energy storage system also provided power to the load.

Especially in August, the sum of the power of wind, solar, and battery (75,712.83 KW h) was less than the load (82,938 KW h), and fuel was used to complete the power supply (7225.17 KW h) to the load.

7.3. Analysis of Emission Reduction

Analysis of cost and benefits may be susceptible to external factors, such as changing fuel purchase prices, the cost fluctuation of transporting fuel and etc. In Antarctica, the analysis of cost and benefits may encounter various complicated situations. The risk of oil spill in transport, atmospheric emissions, or hidden cost of maintenance requirements can rarely achieve comprehensive monetization. However, direct cost savings are in reduced use of fossil fuels. Based on the result of a one-year simulation operation of the standalone renewable energy system in Zhongshan Station, it is found that fuel was used to complete the power supply to the load only in August. The monthly fuel consumption of Zhongshan Station is shown in Table 12.

Table 12. Monthly fuel consumption of Zhongshan Station.

	Jan	Feb	Mar	Apr	May	Jun	Jul	Aug	Sep	Oct	Nov	Dec
Fuel consumption (t)	18.7	18.4	19.31	19.81	21.7	21.5	22.5	21.83	20.46	20.12	17.9	19.1

The existing power supply system of the Antarctic Zhongshan Station has a fuel consumption of 241.33 t per year. The results of simulation operation of the standalone renewable energy system indicate that fuel consumption can be reduced to 2.08 t. The standalone renewable energy system has provided an average annual fuel saving of around 98.8%. Refer to the operation results of Australia's Mawson Station, which is similar in size to Zhongshan Station, Mawson Station used about 0.7 million liters of diesel fuel annually to provide power and heating. Through the introduction of renewable energy supply, the annual fuel saving of Mawson Station was around 32%, equivalent to a saving of 2918 t of carbon dioxide during the first six years of operation [10]. The estimated annual carbon dioxide emissions of existing power supply system of the Antarctic Zhongshan Station were 616.4 t and Zhongshan Station's estimated annual carbon dioxide emissions were 5.3 t in the year of simulation, which can reduced carbon emissions by 611.1 t in one year and 3666.6 t of carbon dioxide during the first six years of operation.

The use of the standalone renewable energy system will improve the health of the local environment and reduce the cost of environmental governance.

7.4. Costs and Benefits of Renewable Energy Applications in Antarctica

Mawson Station has built a wind farm which costs about 8.9 million Australian dollars. The cost of a wind turbine was 0.74 million Australian dollars. Undiscounted simple payback period of the wind farm in Antarctica is estimated to be from 5 to 12 years. For South Pole Station, the project of installing nine 100 KW wind turbines was estimated to cost approximately 4.3 million US dollars. In McMurdo Station, a 1 MW wind turbine has cost 2–3 million US dollars. Net savings of this project remain 1–4 million US dollars over a 20-year life span. Similarly, SANAE IV Station of South Africa has installed a 100 KW turbine with a simple undiscounted payback period of about 10 years. The simple undiscounted payback period of PV system is shorter than wind turbine. For example, a solar thermal system at SANAE IV Station has a payback period of 6 years by saving 10,000 L fuel annually. In the preliminary research stage of the standalone renewable energy system in Zhongshan Station, it is hard to accurately estimate cost savings after the introduction of renewable energy. However, based on the cost savings estimate of other stations, the use of the standalone renewable energy system was estimated to save approximately 1.43 million US dollars in one year.

8. Conclusions

In this study, the standalone renewable energy system used in Zhongshan Station was proposed to achieve long-term stable operation. The meteorological data of Zhongshan Station obtained from a manned weather station in 2015 was comprehensively analyzed. Based on the atmospheric conditions and load data of Zhongshan Station, the physical model, operation principle and mathematical modeling of the proposed power system were designed in this study. The low-temperature performance and characteristics of energy storage system were tested and evaluated. The characteristics of battery and SOC estimation method were also present. To prevent the battery from low-temperature loss of the battery capacity, a temperature control strategy was adopted to keep the storage temperature of the battery above 0 °C. Energy management strategy of the power system was proposed, including a MPPT control strategy for wind turbine and PV array and a power supply strategy. The whole power system is broadly composed of a power generator (wind turbines and PV array), an uploading circuit, a three-phase rectifier bridge, an interleaved Buck circuit, a DC/DC conversion circuit, a switch circuit, a power supply circuit, an amplifier, a driver circuit, a voltage and current monitoring, a load, battery units and a control system. The simulation calculation of power generation of the standalone renewable energy system was presented in this study. Zhongshan Station's estimated annual carbon dioxide emissions were 5.3 t in the year of simulation, which can reduce carbon emissions by 611.1 t in one year and 3666.6 t of carbon dioxide during the first six years of operation. Based on the cost savings estimation of other stations, the use of the standalone renewable energy system was estimated to save approximately 1.43 million US dollars in one year. The results of simulation calculation reveal that the proposed power system can satisfy the power demands of Zhongshan Station in normal operation.

In this study, the proposed power system does not realize completely environmentally friendly operation during its lifecycle because of use of lead–acid batteries and a small amount of fuel. In future work, a more environmentally friendly energy storage system needs to be designed and adopted, such as pumped energy storage, flywheel energy storage, etc. More renewable energy harvesting systems can be used to collect wave energy, temperature and salinity difference energy in Polar Regions. The power generator and the energy storage system need slightly adjusted to achieve 100% environmentally friendly power generation. More works on using advanced sizing methods will be realized to choose power of the photovoltaic array, wind turbine and battery capacity more reasonable.

Author Contributions: Conceptualization, G.Z.; methodology, Y.D.; software, G.Z.; validation, G.Z.; formal analysis, Y.D.; investigation, Y.C.; writing—original draft preparation, G.Z.; funding acquisition, Y.D., X.C.

Funding: This research was funded by the National Key Research and Development Program of China, grant number 2016YFC1400303, 2016YFC1402702.

Acknowledgments: The authors would like to thank the Chinese National Antarctica Research Expeditions for supporting our work in Antarctica. Comments from the anonymous reviewers and the editor are also gratefully appreciated.

Conflicts of Interest: The authors declare no conflict of interest.

References

1. Rothrock, D.A.; Percival, D.B.; Wensnahan, M. The decline in arctic sea-ice thickness: Separating the spatial, annual, and interannual variability in a quarter century of submarine data. *J. Geophys. Res. Oceans* **2008**, *113*, C05003. [CrossRef]
2. Perovich, D.K.; Grenfell, T.C.; Light, B. Transpolar observations of the morphological properties of Arctic sea ice. *J. Geophys. Res.* **2009**, *114*, C00A04. [CrossRef]
3. Haas, C.; Hendricks, S.; Eicken, H. Synoptic airborne thickness surveys reveal state of Arctic sea ice cover. *Geophys. Res. Lett.* **2010**, *37*. [CrossRef]
4. Zuo, G.Y.; Dou, Y.K.; Chang, X.M. Design and Application of a Standalone Hybrid Wind–Solar System for Automatic Observation Systems Used in the Polar Region. *Appl. Sci.* **2018**, *8*, 2736. [CrossRef]
5. Borowy, B.S.; Salameh, Z.M. Optimum photovoltaic array size for a hybrid wind/PV system. *IEEE Trans. Energy Convers.* **1994**, *9*, 482–488. [CrossRef]
6. Deshmukh, M.K.; Deshmukh, S.S. Modeling of hybrid renewable energy systems. *Renew. Sustain. Energy Rev.* **2008**, *12*, 235–249. [CrossRef]
7. Bilal, B.O.; Sambou, V.; Ndiaya, P.A. Optimal design of a hybrid solar–wind–battery system using the minimization of the annualized cost system and the minimization of the loss of power supply probability (LPSP). *Renew. Energy* **2010**, *35*, 2388–2390. [CrossRef]

8. Diaf, S.; Diaf, D.; Belhamel, M. A methodology for optimal sizing of autonomous hybrid PV/wind system. *Energy Policy* **2007**, *35*, 1303–1307. [CrossRef]
9. Dihrab, S.S.; Sopian, K. Electricity generation of hybrid PV/wind systems in Iraq. *Renew. Energy* **2010**, *35*, 1303–1307. [CrossRef]
10. Tin, T.; Sovacool, B.; Blake, D. Energy efficiency and renewable energy under extreme conditions: Case studies from Antarctica. *Renew. Energy* **2010**, *35*, 1715–1723. [CrossRef]
11. Obara, S.; Morizane, Y.; Morel, J. A study of small-scale energy networks of the Japanese Syowa Base in Antarctica by distributed engine generators. *Appl. Energy* **2013**, *111*, 113–128. [CrossRef]
12. Ghoddami, H.; Delghavi, M.B.; Yazdain, A. An integrated wind-photovoltaic-battery system with reduced power-electronic interface and fast control for grid-tied and off-grid applications. *Renew. Energy* **2012**, *45*, 128–137. [CrossRef]
13. Sreeraj, E.S.; Chatterjee, K.; Bandyopadhyay, S. Design of isolated renewable hybrid power systems. *Sol. Energy* **2010**, *84*, 1124–1136. [CrossRef]
14. Gunasekaran, M.; Ismail, H.M.; Chokkalingam, B. Energy Management Strategy for Rural Communities' DC Micro Grid Power System Structure with Maximum Penetration of Renewable Energy Sources. *Appl. Sci.* **2018**, *8*, 585. [CrossRef]
15. Vasiliev, M.; Alameh, K.; Nur-E-Alam, M. Spectrally-Selective Energy-Harvesting Solar Windows for Public Infrastructure Applications. *Appl. Sci.* **2018**, *8*, 849. [CrossRef]
16. Akinyele, D.; Belikov, J.; Levron, Y. Battery Storage Technologies for Electrical Applications: Impact in Stand-Alone Photovoltaic Systems. *Energies* **2017**, *10*, 1760. [CrossRef]
17. Carroquino, J.; Bernal-Agustín, J.; Dufo-López, R. Standalone Renewable Energy and Hydrogen in an Agricultural Context: A Demonstrative Case. *Sustainability* **2019**, *11*, 951. [CrossRef]
18. Gabrielli, P.; Gazzani, M.; Martelli, E. Optimal design of multi-energy systems with seasonal storage. *Appl. Energy* **2018**, *219*, 408–424. [CrossRef]
19. Ma, T.; Yang, H.X.; Lu, L. Technical feasibility study on a standalone hybrid solar-wind system with pumped hydro storage for a remote island in Hong Kong. *Renew. Energy* **2014**, *69*, 7–15. [CrossRef]
20. Argyrou, M.; Christodoulides, P.; Kalogirou, S. Energy storage for electricity generation and related processes: Technologies appraisal and grid scale applications. *Renew. Sustain. Energy Rev.* **2018**, *94*, 804–821. [CrossRef]
21. Wang, Y.J.; Zhang, C.B.; Chen, Z.H. A method for joint estimation of state-of-charge and available energy of LiFePO$_4$ batteries. *Appl. Energy* **2014**, *135*, 81–88. [CrossRef]
22. Dong, G.Z.; Chen, Z.H.; Wei, J.W. An online model-based method for state of energy estimation of lithium-ion batteries using dual filters. *J. Power Sources* **2016**, *301*, 277–286. [CrossRef]

© 2019 by the authors. Licensee MDPI, Basel, Switzerland. This article is an open access article distributed under the terms and conditions of the Creative Commons Attribution (CC BY) license (http://creativecommons.org/licenses/by/4.0/).

Review

Energy Management in Microgrids with Renewable Energy Sources: A Literature Review

Yimy E. García Vera [1], Rodolfo Dufo-López [2,*] and José L. Bernal-Agustín [2]

1. Electronic Engineering, San Buenaventura University, Bogotá 20, Colombia; yegarcia@usbbog.edu.co
2. Electrical Engineering Department, University of Zaragoza, 50018 Zaragoza, Spain; jlbernal@unizar.es
* Correspondence: rdufo@unizar.es

Received: 24 July 2019; Accepted: 11 September 2019; Published: 13 September 2019

Abstract: Renewable energy sources have emerged as an alternative to meet the growing demand for energy, mitigate climate change, and contribute to sustainable development. The integration of these systems is carried out in a distributed manner via microgrid systems; this provides a set of technological solutions that allows information exchange between the consumers and the distributed generation centers, which implies that they need to be managed optimally. Energy management in microgrids is defined as an information and control system that provides the necessary functionality, which ensures that both the generation and distribution systems supply energy at minimal operational costs. This paper presents a literature review of energy management in microgrid systems using renewable energies, along with a comparative analysis of the different optimization objectives, constraints, solution approaches, and simulation tools applied to both the interconnected and isolated microgrids. To manage the intermittent nature of renewable energy, energy storage technology is considered to be an attractive option due to increased technological maturity, energy density, and capability of providing grid services such as frequency response. Finally, future directions on predictive modeling mainly for energy storage systems are also proposed.

Keywords: microgrids; energy management; renewable energy; optimization; photovoltaic; energy storage

1. Introduction

The exponential demand for energy has led to the depletion of fossil fuels such as petroleum, oil, and carbon. This, in turn, increases the greenhouse effect gases. Energy systems have incorporated small-scale and large-scale renewable sources such as solar, wind, biomass, and tidal energy to mitigate the aforementioned problems on a global scale [1]. Global energy demand will grow by more than a quarter to 2040, when renewable sources are expected to represent 40 percent of the global energy mix. The reliability of the renewable sources is a major challenge due mainly to mismatch between energy demand and supply [2]. Renewable energy resources, distributed generation (DG), energy storage systems, and microgrids (MG) are the common concepts discussed in several papers [3]. The increase in the demand for energy and the rethinking of power systems has led to energy being generated near the places of consumption. This energy is derived from renewable sources, which are becoming increasingly competitive due to a drop in prices, especially in the case of photovoltaic solar and wind energies [4].

Due to strong dependency on climatic and meteorological conditions, in many cases the optimal system is a hybrid renewable energy system (considering one or more renewable sources) with battery storage systems (and in some cases including diesel generator) [5]. The hybrid energy systems are typically used for electricity supply for several applications such as houses or farms in rural areas without grid extension, telecommunication antennas, and equipment, and many other stand-alone systems [6,7]. In many cases these hybrid systems imply the highest reliability and lowest costs compared to systems with only one energy source [8,9].

A microgrid consists of a set of loads, energy storage equipment, and small-scale generation systems [10]. It can be defined in a broader sense as a medium or low distribution grid, which has distributed generation including renewable and conventional sources (hybrid systems) with storage units that supply electrical energy to the end users. The reliability of the microgrid is improved by the storage and it is used to complement the intermittency of the PV and wind output power [11–13]. These microgrids have communication systems that are necessary for real time management [14]. Microgrids can also operate either in isolation or when connected to a grid [15]. Based on the type of source they manage, microgrids can be classified as direct current line (DC), alternating current line (AC), or hybrid (shown in Figure 1).

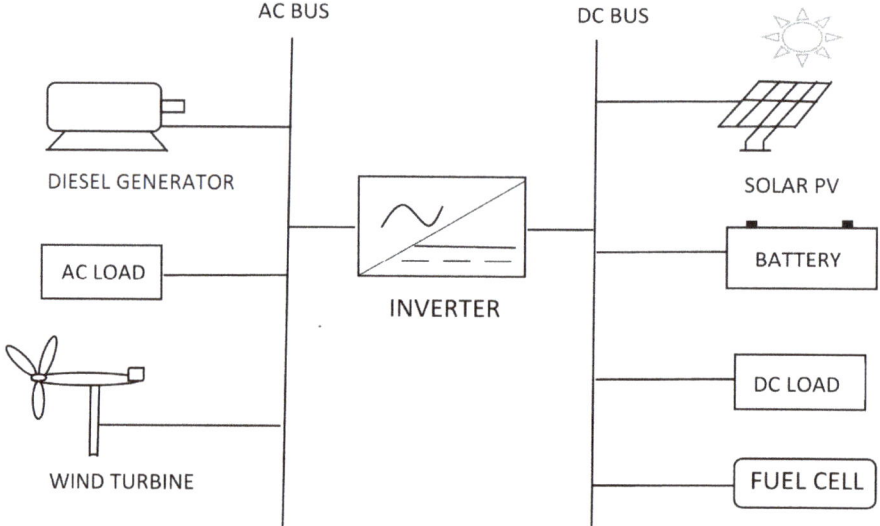

Figure 1. A hybrid isolated microgrid scheme.

In a microgrid, it is essential to maintain the power supply-demand balance for stability because the generation of the intermittent distributed sources such as photovoltaic and wind turbines is difficult to predict and their generation may fluctuate significantly depending on the availability of the primary sources (solar irradiation and wind). The supply-demand balancing problem becomes even more important when the microgrid is operating in stand-alone mode where only limited supply is available to balance the demand [16]. Energy management optimization in microgrids is usually considered as an offline optimization problem [17].

Microgrids supported with renewable energies can be classified as smartgrids, which provide a set of technological solutions to allow information exchange between the consumers and the distributed generation. An energy management system (EMS) is defined as an information system, which provides the necessary functionality when supported on a platform to ensure that generation, transmission, and distribution supply energy at minimal cost [18]. Energy management in the microgrids involves a control software that permits the optimal operation of the system [19]. This is achieved by considering the minimal required cost and two microgrid operation modes (isolated and interconnected). The variability of resources such as solar irradiation and wind speed must be accounted for when considering microgrids with renewable energy sources [20].

A review on the studies related to the energy management of microgrids can be found in [21]. A few authors have solved the problem of energy management using different techniques to achieve an optimal microgrid operation. However, these techniques must incorporate better solution strategies due to the integration of distributed generation, storage elements, and electric vehicles.

Other recent papers [22] have reviewed various integration methods for renewable energy systems based on storage and demand response. This covers two main areas, namely (1) the optimal usage of storage, and (2) improvement of user participation via demand response mechanisms and other collaborative methods. The authors in [23] reviewed energy management strategies for hybrid renewable energies. The above review covered different configurations of stand-alone and grid-connected hybrid systems. Other review papers [24] have shown the control objectives of the microgrid supervisory controllers (MGSC) and energy management systems (EMS) for microgrids. Table 1 shows the contributions of the review papers related to the energy management of microgrids. Unlike the cited papers, this paper focuses on the incorporation of better strategies for the control of energy (both heat and electrical) flow between the hybrid system sources and load. Furthermore methods of energy management in stand-alone hybrid microgrid considering the battery degradation are also discussed.

Table 1. Microgrids energy management review papers.

Reference	Contributions
[21]	Authors presented a comparative analysis on decision making strategies for microgrid energy management systems. These methods are selected based on their suitability, practicability, and tractability, for optimal operation of microgrids.
[22]	Energy management integration methods, demand response, and storage systems are reviewed. Authors used more accurate models for storage including key factors such as the derating factors due temperature charge/discharge rate and ageing.
[23]	Authors presented a review on strategies and approaches used to implement energy management in stand-alone and grid-connected hybrid renewable energy systems.
[24]	Authors showed an extensive review on energy management methodologies applied in microgrids. EMS for real-time power regulation and short-/long-term energy management are reviewed.
[25]	Authors showed previous solutions approaches, optimization techniques, and tools used to solve energy management problem in microgrids. It includes heuristic, agent-based, MPC, evolutionary algorithms, and other methods.
[26]	Authors showed an overview of the latest research developments using optimization algorithms in microgrid planning and planning methodologies.
[27]	Authors presented an overview of current hybrid microgrids and optimization methods and applications.
[28]	Authors showed in detail the optimization of distributed energy microgrids in both the grid-connected and stand-alone mode.

2. Microgrid Optimization Techniques

Energy management of a microgrid involves a comprehensive automated system that is primarily aimed at achieving optimal resource scheduling [25–27]. It is based on advanced information technology and can optimize the management of distributed energy sources and energy storage system [28]. The microgrid optimization problem typically involves the following objectives:

Maximize the output power of the generators at a particular time;
Minimize the operating costs of the microgrid;
Maximize the lifetime of energy storage systems;
Minimize the environmental costs.

Some of the classic optimization methods include mixed integer linear and non-linear programming. The objective function and constraints used in linear programming are linear functions with real-valued and whole-valued decision variables. Dynamic programming methods are used to solve more complex problems that can be discretized and sequenced. The problem is typically broken down into sub-problems that are optimally solved. Then, these solutions are superimposed to develop an optimal solution for the original problem.

Metaheuristics is another important alternative in microgrid optimization. Heuristic techniques are combined to approximate the best solution using genetic algorithms, biological evolution, and statistical mechanisms for achieving optimal operation and control of microgrid energy.

Predictive control techniques are used in applications where predicting the generation and loading is necessary to guarantee effective management of stored energy. This typically combines stochastic programming and control. The most remarkable among these techniques are the ones to predict the deterioration of elements of the grid, mainly storage systems.

Optimization methods based on a multi-agent used on microgrids allow a decentralized management of the microgrid and consist of sections having autonomous behavior to execute the tasks with defined objectives. These agents, which include loads, distributed generators and storage systems, communicate with each other to achieve a minimal cost.

Stochastic methods and robust programming are used to solve the optimization functions when the parameters have random variables, particularly in artificial neural networks, fuzzy logic, and game theory.

A few more methods can be derived from a combination of the aforementioned techniques such as stochastic and heuristic methods and enumeration algorithms.

3. Microgrid Energy Management with Renewable Energy Generation

A microgrid is composed of different distributed generation resources that are connected to the utility grid via a common point. Figure 2 shows a microgrid energy management mode along with several features that are modules of human machine interfaces (HMI), control and data acquisition, load forecast, optimization, etc. [29].

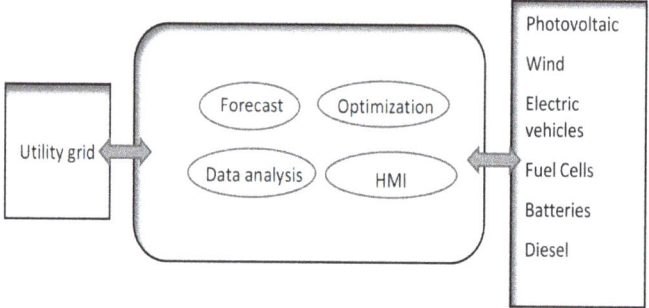

Figure 2. Microgrid energy management [29].

Many researchers have addressed energy management by implementing different approaches. However, all approaches have focused on determining the most optimal and efficient microgrid operation. The following sub-sections discuss and classify these strategies and solutions.

3.1. Energy Management Based on Linear and Non-Linear Programming Methods

Ahmad et al. [30] presented a technical and economic method to optimize a MG based on mixed integer linear programming (MILP). This paper presents the advantages of programming the generation of distributed sources, managing the intermittency and volatility of this type of generation, and reducing load peaks. The cost function is solved via linear programming based on a general algebraic modeling system (GAMS). Simulations to optimize MG size are performed via software called HOMER. Taha and Yasser [31] presented a robust algorithm based on a predictive control model for an isolated MG. The model incorporates multi-objective optimization with MILP, which minimizes the cost, energy consumption, and gas emission due to diesel generation in the MG.

Sukumar et al. [32] proposed a mixed method for MG energy management. This was achieved by combining the utility grid and fuel cell power. The problem is solved using linear optimization methods, and the on/off states of the utility grid are solved via MILP. A particle swarm optimization (PSO) method was used to obtain an optimal energy storage system size.

Tim et al. [33] proposed a system for energy management in an interconnected MG that adopted a centralized approach based on the concept of flexibility for the final users. An optimal economic dispatch was obtained using quadratic programming. This grid was integrated with a photovoltaic system and the constraints must satisfy the demand. The algorithm was tested on an IEEE 33 node modified grid.

Delgado and Domínguez-Navarro [34] presented an algorithm based on linear programming for MG energy management that allowed the optimal operation of either generators or controllable and non-controllable loads. The optimization problem involves the optimal dispatch of generators (diesel) while meeting the operational and economic constraints imposed by the purchase and sale of energy corresponding to each component (generators, storage systems, and loads).

Helal et al. [35] analyzed an energy management system for a hybrid AC/DC MG in an isolated community that employs a photovoltaic system for desalination. The proposed optimization algorithm was based on the mixed integer non-linear programming, wherein the objective function minimizes the daily operating costs.

Umeozor and Trifkovic [36] researched the energy management of a MG based on MILP via the parametrization of the uncertainty of solar and wind energy generation in the MG. The optimization is achieved at two levels. First, the parametrization scheme is selected; second, the operational decisions are made the problem considers the variation in market prices and the disposition of the storage systems.

Xing et al. [37] presented an energy management system based on multiple time-scales. The optimization problem considers two aspects: A diary static programming and dynamic compensation in real time. This is solved via a mixed-integer quadratic programming method using optimal load flows, and the load state of the batteries are predicted using wind and solar radiation data.

Correa et al. [38] proposed an energy management system based on a virtual power plant (VPP). The studied MG has solar panels and storage systems and works in an interconnected manner. These elements are programmed/modeled using linear programming methods to minimize the operating costs. Renewable energies are incorporated into an energetic model, similar to the Colombian one, and are mainly based on hydric resources.

Cardoso et al. [39] analyzed a new model to observe the battery degradation of a MG. The problem is solved using stochastic mixed-integer linear programming, taking several factors such as loads and different sources of energy generation, costs, constraints, grid topology, and local fees for energy into consideration.

Behzadi and Niasati [40] analyzed a hybrid system that consists of a photovoltaic (PV) system, battery, and fuel cells. Performance analysis was conducted using the TRNSYS software, and the sizing was determined either using the genetic algorithm in the HOGA software (now called iHOGA), manual calculations, or the HOMER software. Three energy management strategies were tested for energy dispatch in this hybrid system. The excess energy was checked in each system and a decision was taken to either produce hydrogen or charge the battery or both.

3.2. Energy Management Based on Metaheuristic Methods

Dufo-López et al. [41] proposed a control strategy for the optimal energy management of a hybrid system based on genetic algorithms. The system is composed of renewable sources (PV, wind, and hydro), an AC generator, electrolyzer, and fuel cells. Energy management is optimized to minimize the operating costs, which enables the use of the excess energy generated by the renewable sources to charge the batteries or produce hydrogen in the electrolyzer. The load that cannot be supplied by the renewable sources can be obtained by either discharging the battery or using fuel cells.

Das et al. [42] studied the effect of adding internal combustion engines and gas turbines to a stand-alone hybrid MG with photovoltaic modules. A multi-objective genetic algorithm was used to optimize this system based on the energy costs and overall efficiency. Two strategies, both electric and thermal, were used to track the load. All the analyzed systems satisfied the electrical demand when combined with both heating and cooling.

Luna et al. [43] presented an energy management system that operates in real time. Three cases were studied considering the perfect, imperfect, and exact predictions. The employed optimization model was tested in both a connected and an isolated MG, with large imbalances between the generation and load.

An economic dispatch and battery degradation model has been proposed in [44], wherein genetic algorithms were used for energy supply options via a diesel generator. The results showed that an increase in the battery lifespan decreases the operational costs of a MG. This method was validated in a hybrid MG composed of a diesel generator and photovoltaic system.

Chaouachi et al. [45] proposed a multi-objective, intelligent energy management system for a MG that minimizes the operational costs and environmental impact. An artificial neural network has been developed to predict the photovoltaic and wind power generation 24 and 1 h in advance, respectively, along with the load demand. The multi-objective intelligent energy management system is composed of multi-objective linear programming. The battery scheduling is obtained using a fuzzy logic-based expert system.

Li et al. [46] presented a study on MG optimization based on the particle swarm algorithm that can operate a connected or isolated MG. The proposed approach considers the fluctuations in the renewable sources and load demands in the MG, with appropriate advance (24 h) forecasts available to overcome these fluctuations.

Nivedha et al. [47] analyzed a MG containing/supporting wind power generation, fuel cells, a diesel generator, and an electrolyzer. A fuel cell is used when the energy demand is not covered by the wind turbine, to ensure energy balance when operating diesel generators to reduce the operational costs. The fuel cell operates to meet the high load demand, resulting in economic MG operation with a ~70% cost saving using the particle swarm optimization algorithm.

Abedini et al. [48] presented an energy management system for a photovoltaic/wind/diesel stand-alone hybrid MG, which is optimized using a particle swarm algorithm with Gaussian mutation. This study minimizes both the capital and fuel costs of the system.

Nikmehr et al. [49] studied an optimal generation algorithm applied to a MG based on optimization via the imperial competitive algorithm. This algorithm solves the load uncertainty and distributed generators, along with the economic dispatch of the generating units. This algorithm is comparable to methods such as the Monte Carlo method, and has been tested in interconnected MGs.

Marzband et al. [50] presented an energy management system for an isolated MG using the artificial bee colony algorithm (ABC). A stochastic approach is required to analyze the economic dispatch of the generating units inside a MG, given the intermittent nature of solar energy resources and wind generation. The results showed a 30% decrease in costs. The non-dispatchable generation and load uncertainty are managed using neural networks and Markov chains.

Kuitaba et al. [51] presented a new method to optimize an interconnected MG, which combines an expert system based on fuzzy logic and a metaheuristic algorithm known as Grey Wolf optimization. This method involves minimizing both the costs of the generating units and the emission levels of the fossil fuel sources. This method lowers MG costs by considering the optimal capacity of the batteries and reducing the consumption of fossil fuels.

Papari et al. [52] analyzed energy management in a MG connected to a direct current utility grid. The optimization is implemented using the crow search algorithm (CSA), which is a metaheuristic optimization method that imitates the behavior of a crow to store and hide food.

Wasilewski [53] presented a metaheuristic optimization method to optimize a MG. The methods include the evolutionary and particle swarm algorithms. These methods account for the fact that the deterministic conditions assumed in the problem impose an important limit on the employed methodology. However, it also recognizes the uncertainty of using renewable energies.

Ogunjuyigbe et al. [54] presented a technique based on a genetic algorithm for the optimal location of both renewable generation and batteries in a stand-alone MG. The proposed multi-objectives are to reduce operational and life cycle costs, and dump energy. The optimization allows variations in the radiation and wind sources, and extracts data from a load profile to optimize the MG.

Kumar and Saravanan [55] proposed an algorithm based on the demand prediction over 24 h in a MG using the artificial fish swarm optimization method. Thus, the demand can be planned in advance, considering both renewable and non-renewable generation. The algorithm is used to program the sources, load, and storage elements. They system includes a wind turbine, two photovoltaic generators, a fuel cell, a micro-turbine, and a diesel generator.

A particle swarm algorithm has been proposed in a recent paper by Hossain et al. [56] for energy management in a grid-connected MG. A model for charging and discharging a battery has been formulated. The proposed cost function reduces costs by 12% over a total time horizon/period of 96 h, with time intervals of one hour. These results can be adjusted in real time.

Azaza and Wallin [57] studied energy management in a MG with a hybrid system consisting of wind turbines, photovoltaic panels, diesel generator, and battery storage. A multi-objective particle swarm optimization is used, which evaluates the probability of losing energy supply over a time horizon/period of 6 months each during summer and winter.

Motevasel and Seifi [58] presented an expert system for energy management (EEMS) in a MG that contains wind turbines and photovoltaic generation. Neural networks are used to predict wind turbine generation. The bacterial foraging algorithm is used for the optimization, while the optimization of the multi-objective problem is obtained by the EEMS module by applying an improved bacterial foraging-based fuzzy satisfactory algorithm.

Rouholamini and Mohammadian [59] proposed optimal energy management for a grid-connected hybrid generation system, including PV generator, wind turbine, fuel cell, and electrolyzer. This system trades power with the local grid using real time electricity pricing over a 24-h time horizon/period based on the simulation results. The interior search algorithm was used to optimize the energy management in the above case.

3.3. Energy Management Based on Dynamic Programming Techniques

Shuai et al. [60] proposed an energy management system for a MG based on dynamic programming and mixed-integer non-linear programming optimization. The MG is interconnected to the grid and decisions are made using the Bellman equation. Historical data are used off-line, while considering the power flow and battery storage as constraints. Using the algorithm in multiple MGs simultaneously is a feasible possibility.

Almada et al. [61] proposed a centralized system for energy management of a MG either in the stand-alone or interconnected modes. In the stand-alone mode, the fuel cell only works if the battery is less than 80%. In the interconnected mode, a 60% threshold is required to ensure reliable behavior.

Wu et al. [62] proposed an algorithm based on dynamic programming for the management and control of stand-alone MGs. The deep learning algorithm works in real time, which permits intra-day scheduling to obtain a control strategy for MG optimization, while sending information from local controllers within the framework of centralized management.

Zhuo [63] proposed an energy management system using dynamic programming to manage a MG with renewable generation sources and batteries. The objective was to maximize the benefits from the sale of renewable energy and minimize the cost required to satisfy the energy demand. The author used a non-regulated energy market where electricity prices fluctuate and the battery control actions are determined by dynamic programming.

Choudar et al. [64] presented an energy management model based on the battery state of charge and ultra-capacitors. The hierarchic structure of optimal MG management has four states or operating modes: Normal operating mode, photovoltaic limitation mode, recovering, and stand-alone modes.

Marabet et al. [65] proposed an energy management system for a laboratory scaled hybrid MG with wind, photovoltaic, and battery energy. The control and data acquisition system are operated in real time. The energy management system is based on a set of rules, and optimizes the MG performance by controlling and supervising the power generation, load, and storage elements.

Luu et al. [66] presented a dynamic programming method and methodology based on the rules applied to a stand-alone MG containing diesel and photovoltaic generators, and a battery. The constraints are governed by the power balance between generation and consumption, along with the capacity of each distributed generator. Dynamic programming is used to minimize the operational and emission costs. The constraints are the power balance between offer and demand, along with the operating capacity of each distributed generator.

3.4. Energy Management Based on Multi-Agent Systems

Boudoudouh and Maâroufi [67] proposed an energy management system in a MG with renewable energy sources. Simulations were run using the Matlab-Simulink and java platform for agent developers (JADE) software. The reliability of this model was validated by fulfilling requirements such as autonomy and adaptability in the MG management system with load variation.

Raju et al. [68] studied energy management in a grid outage divided into two MGs, which contains two photovoltaic and wind generators each and a local load. A multi-agent management system based on the differential evolution algorithm in JADE was used to minimize the generation costs from the intermittent nature of the solar resource and randomness of load. This system also addressed the price variation in the grid, and the critical loads were considered while selecting the best solution.

Bogaraj and Kanakaraj [69] presented an energy management proposal based on intelligent multi-agents for a stand-alone MG, which maintains the energetic balance between the loads, distributed generators, and batteries. The agents consist of photovoltaic systems, wind turbines, fuel cells, and battery banks. Loads are divided in three groups based on their priority. The auto-regressive moving average models (ARMA) were used to predict the generation. Cases covering high and low irradiation, and low wind were analyzed. The system used a dynamic compensator to balance the reactive power.

Anvari-Moghaddam et al. [70] presented an energy management system for a microgrid that includes houses and buildings. The optimization process for the energy management system involves the coordination of management in distributed generation (DG) and response to the demand. The main objectives of the cost function are to minimize the operating costs and meet the thermic and electrical needs of the clients. The communication platform used by the agents is based on the hypertext (HTPP) communication protocol.

In the study investigated in [71], Nunna and Doolla used an energy management system based on multi-agents, which considers different types of load patterns and the energy available from the distributed energetic resources. They proposed a novel mechanism that encouraged clients to participate. This proposal was validated in interconnected grids using the JADE programming language. The management system reduces the consumption peaks and offers the clients an attraction benefit–cost ratio.

Dou and Liu [72] presented a decentralized multi-objective hierarchical system based on the agents in an interconnected smart MG, minimizing the operating and emission costs and line losses.

The authors in [73] researched decentralized energy management based on the multi-agents contained in a MG, using cognitive maps with fuzzy logic. The intelligent agents refer to the distributed generators, batteries, electrolyzer, and fuel cells. Centralized and decentralized approaches were compared and it showed that the decentralized approach offers the advantage of partial operation under certain circumstances such as during a system malfunction or failure.

Mao et al. [74] presented a hybrid energy management system for a MG based on multi-agents, which incorporates both the centralized and decentralized approaches and optimizes the economic operation of the MG. A novel simulation platform for energy management systems was designed based on the client-server framework and implemented in the C++ environment.

Netto et al. [75] developed a real time framework for energy management in a smart MG in the islanded mode using a multi-agent system. The RSCAD software was used to simulate the MG using the TCP/IP protocol for the purposes of testing and real time operation.

3.5. Energy Management in Microgrids Based on Stochastic Methods and Robust Programming

Che Hu et al. [76] showed an energy management model for a MG wherein the uncertainty in the supply and energy demand are taken into account. Uncertainty in wind and photovoltaic generation, and demanded energy is considered. The stochastic programming of two states was formulated using the GAMS and was tested on a real grid at the Nuclear Energy Research Centre in Taiwan. The battery capacity was optimized in the first stage, while an optimal operation strategy for the MG was evaluated in the second stage.

The author in [77] presented an optimization system for a hybrid MG using a multi-objective stochastic technique. The objective function presented in this study minimizes the system losses and reduces the operating cost of the renewable resources, which were used at different points of the MG. The problem was formulated using the weighting sum for the total operating cost and losses of the feeding systems. The proposed scheme was solved using mixed integer linear programming and tested on the IEEE 37 node distribution system.

Lu et al. [78] proposed a dynamic pricing mechanism that achieves an optimal operating performance. This mechanism was applied to a grid composed of multiple MGs, to evaluate the uncertainty of renewable energy integration on a large scale. An optimization scheme was developed at two levels: The pricing mechanism guaranteed the market operator's energy operation in the upper level, while in the lower level the MG transactions were developed.

Xiang et al. [79] proposed an optimization model for an interconnected MG based on a model using the Taguchi orthogonal matrices. The uncertainty in the renewable energy and load demand were determined by an interval based on error prediction.

Hu et al. [80] introduced an optimization method for an interconnected grid that is divided in two stages. A conventional generator is used in the first stage, while the second stage ensures an economical dispatch of the conventional and distributed generation using hourly marketing. This combination permits management of the uncertainty in renewable generation using the Lyapunov optimization method.

Shen et al. [81] presented a stochastic energy management model for an interconnected MG. The uncertainty level is managed using Latin hypercube sampling based on the Monte Carlo method, which generates various scenarios for the distributed resources, load, and electricity price. A sensitivity analysis is performed to determine the standard deviation of the expected price and level of reliability.

Rezai and Kalantar [82] proposed a stochastic energy management system for a stand-alone MG based on the minimization of frequency deviations. Operating costs of the MG include conventional and distributed generation, and reserves and incentives for generation using renewable sources. The outputs of the conventional generators were also analyzed for various contingencies to demonstrate the robustness of the proposed approach.

Su et al. [83] studied a model for the efficient programming of an interconnected MG, which minimizes the operating costs of the conventional generators, battery degradation, and commercial costs corresponding to the energy from the utility grid. This model follows two stages.

The first stage involves optimization of the MG, while the second stage involves analysis of the power output to calculate the MG energy losses in real time.

Farzin et al. [84] proposed an energy management system for an isolated MG. The islanding event was treated as a normal probability distribution of the failures in the utility grid. The objective was to minimize the MG operating costs. This included costs associated with the microturbine operation, wind turbines, batteries, and load disconnection.

Liu et al. [85] proposed an energy management system for an interconnected MG considering renewable energies and load uncertainties. The energy management is divided in two sub-problems: The first involves scheduling within the defined energy boundaries for system protection, while the second evaluates the real time energy capacity deviation limit for frequency regulation. The presented approach was found to be more cost effective.

Kuztnesova et al. [86] proposed a decentralized energy management system for an interconnected MG using agent-based modeling and robust optimization. The MG performance was evaluated in terms of the cost from the power imbalances associated with the uncertainty of renewable generation and load power demand.

Zachar and Daoutidis [87] proposed a hierarchic control mechanism to regulate and supervise the loads and dispatchable energy inside a MG. Stochastic optimization was used on a low scale to avoid errors in the forecast of renewable energies. Deterministic optimization was realized on a fast scale to update the optimal dispatch conditions.

Battistelli et al. [88] proposed an energy management system for a remote hybrid AC/DC MG, which ensures economical dispatch in spite of the uncertainties associated with the use renewable energy sources. A load control is determined (thermic and electric vehicles) based on the demand, while taking the limits of the generators, controllable loads, and charge and discharge of batteries into consideration.

Lujano et al. [89] developed an optimal load management method for hybrid systems composed by the wind tubine, battery bank, and diesel generator. The autoregressive moving average (ARMA) was used to predict the wind speed.

The results showed that the load management strategy improved wind power usage by shifting the controllable loads to the wind power peaks, thus increasing the charge in the battery bank. This research contributed strategies for the energy management of hybrid MGs.

3.6. Energy Management Using Predictive Control Methods

Zhai et al. [90] proposed a predictive robust control that can be applied to a stand-alone MG. The management model employed mixed integer programming. The MG is composed of wind and PV generators, batteries, and loads.

Zhang et al. [91] presented a model predictive control (MPC) method to manage a MG that integrates both distributed and renewable generation. The model's objective is to reduce the costs and constraints in both generation and energy demand.

Minchala Ávila et al. [92] proposed a methodology based on predictive control for energy management in a stand-alone MG. The controller operates the battery energy in a centralized manner and performs a load elimination strategy to ensure balance in the MG power output.

Ju et al. [93] investigated an energy management system for a hybrid MG taking the degradation costs of the energy storage systems into consideration. The proposal consists of a two-layer predictive control for the hybrid MGs, which use batteries and supercapacitors as storage systems. An important contribution of this work is that the degradation costs of the supercapacitors and batteries were modeled, which allows more accurate assessment of the MG operating costs.

Valencia et al. [94] proposed an energy management model for a MG that uses predictive control, which involves the prediction of the intervals using fuzzy logic. This allows the representation of the non-linearity and dynamic behavior of the renewable sources.

Genesan et al. [95] presented an energy management system for a MG based on a control algorithm to integrate and manage various types of generation such as the PV, distributed generation, energy storage systems, and UPS from the supply grid and different loads. The transition problem between the storage systems and PV generation is solved via control and communication, which functions on a TCP/IP protocol.

García Torres and Bordons [96] introduced optimal programming in a hybrid MG, based on a predictive control model that is solved using mixed integer quadratic programming. They integrated the operating costs and MG optimization, which includes the degradation costs of all the components of the hybrid system, mainly the hydrogen-based storage systems.

Solanki et al. [97] presented a mathematical model of the smart loads and energy management of a stand-alone MG. Loads are modeled using neural networks. Energy management is realized with the predictive control method, which performs an optimal power dispatch taking the elements and controllable loads into consideration.

Oh et al. [98] proposed a multi-step predictive control model for a MG over a time horizon/period of 180 min in 15 min steps. This includes conventional and renewable energy generators, energy storage elements, and both critical and non-critical loads. The cost function was formulated considering the costs associated with fuel consumption, renewable energy reduction, battery state of charge, and amount of load shedding.

A proposal has been presented by Prodan et al. [99] for the energy management of a MG based on a fault-tolerant predictive control design. One of their many contributions includes the extension of the useful battery life by decreasing the charge and discharge cycles.

Wu et al. [100] presented an optimal solution for the operation of a hybrid system using solar energy and battery storage. The battery plays a significant role in the storage of grid power during off-peak periods and supply of power to the customers during peak demand. Thus, scheduling the hybrid system leads to the minimal power consumption from the grid and reduces a customer's monthly cost.

Thirugnanam et al. [11] proposed a battery strategy management. The main objective tries to reduce the fuel consumption in DG, reduce fluctuating PV power, and control the battery charge and/or discharge rate to improve the battery life cycle. The battery charge/discharge rate control model considers the battery SOC limits, wherein the batteries are not charged or discharged beyond the specified limits.

Dufo-López et al. [101] presented a technique to optimize the daily operation of a diesel-wind-PV hybrid, using MPC with forecast data of the irradiation, wind speed, temperature, and daily load. The main contribution of this work is daily optimization that accounts for the degradation of the lead-acid battery by corrosion and capacity losses, using the advanced model presented by Schiffer et al. [102]. This parameter is important when considering the operating costs of the MG, as the useful life and replacement of the batteries can be estimated more accurately. The optimization is executed using genetic algorithms.

3.7. Energy Management Based on Artificial Intelligence Techniques

Elseid et al. [103] defined the role of energy management in a MG as a system that autonomously performs the hourly optimal dispatch of the micro and utility grids (when interconnected) to meet the energy demand. In the above study, the authors used a CPLEX algorithm developed by IBM.

Mondal et al. [104] proposed an energy management model for a smart MG based on game theory, using a distributed energy management model. In this scheme, the MG selects a strategy to maximize its benefits with respect to the cost and adequate use of energy.

Prathyush and Jasmín [105] proposed an energy management system for a MG using a fuzzy logic controller that employs 25 rules. The main objective is to decrease the grid power deviation, while preserving the battery state of charge.

Leonori et al. [106] proposed an adaptive neural fuzzy inference system using an echo state network as a predictor. The objective was to maximize the income generated from energy exchange with the grid. The results showed that the energy management performance improved by 30% over a 10 h prediction horizon/period.

De Santis et al. [107] introduced an energy management system for an interconnected MG using fuzzy logic based on the Mamdani algorithm. The main objective is to take decisions on the management tasks of the energy flow in the MG model, which is composed of renewable energy sources and energy storage elements. The optimization was realized in a scheme that combines fuzzy logic and generic algorithms.

Venayagamoorthy et al. [108] proposed an energy management model for a MG connected to the main power grid. The MG maximizes the use of renewable energies and minimizes carbon emissions, which makes it self-sustainable. The management system is modeled using evolutionary adaptive dynamic programming and learning concepts using two neural networks. One of the neural networks is used for the management strategy, while the other used to check for an optimal performance. The performance index is evaluated in terms of the battery life, use of renewable energy, and minimization of the controllable load.

Ma et al. [109] proposed an algorithm using game theory based on the leaders and followers for energy management. This approach aims at maximizing the benefits available to active consumers of the MG, while keeping the Stackelberg balance to ensure an optimal distribution of benefits.

Jia et al. [110] formulated an adaptive intelligence technique for the energy management of an interconnected MG, which uses energy storage elements. The objective is to minimize any load fluctuations due to uncertainties in the renewable energy generation. The load profile is managed by storage elements and ultra-capacitors.

Arcos-Avilés et al. [111] presented an energy management algorithm based on low-complexity fuzzy logic control for a residential grid-connected MG, which includes renewable distributed generation and batteries.

Aldaouab et al. [112] proposed an optimization method using genetic algorithms for residential and commercial MGs. The MG uses PV-solar energy, microturbines, a diesel generator, and an energy storage system.

Liu et al. [113] proposed a Stackelberg game approach for energy management in a MG. A management system model that takes the fee for the PV energy into account was introduced, which includes the profits from the MG operator and a utility model for the PV consumers.

Nnamdi and Xiaohua [114] proposed program consisting on an incentive-based demand response for the operations of the grid connected MG. The game theory based demand response program (GTDR) was used to investigate the grid connected operational mode of a MG. The results showed that lower costs could be achieved in the MG when the DG benefit of the grid operator is maximized at the expense of minimizing the fuel/transaction costs.

3.8. Energy Management Based on Other Miscellaneous Techniques

Astaneh et al. [115] proposed an optimization scheme to find the most economic configuration for a stand-alone MG, which has a storage system with lithium batteries, and considered different control strategies for energy management. The lifetime of lithium batteries is estimated using an advanced model based on electrochemistry to evaluate the battery longevity and its lifetime.

Neves et al. [116] presented a comparative study on the different objectives of the optimization techniques for the management of stand-alone MGs. This approach is primarily based on linear programming and genetic algorithms. The results showed that the optimization of the controllable loads could result in an operating cost reduction and inclusion of renewable energies.

Wei et al. [117] proposed an iterative and adaptive algorithm based on dynamic programming to enable optimal energy management and control a residential MG. The charge/discharge level of the battery is treated as a discreet problem in hourly steps. The decisions on the energy supply for a residential load with respect to the energy fee are made in real time.

Yan et al. [118] studied the design and optimization of a MG using a combination of techniques such as mixed integer programming for the optimization of energy management, and the probabilistic Markov model to represent the uncertainty of PV generation. The design included a linear model to evaluate the MG lifetime.

Akter et al. [119] proposed a hierarchic energy management model for an interconnected residential MG serving prosumers, which includes a local control mechanism that shares information with a central controller for energy management.

In the research presented in [120], an energy management system was designed for a hybrid system combining wind, PV, and diesel generation. The system operates both on- and off-grid. Thus, there exists a control mechanism within the inverter for transfers between the micro and utility grids.

Lai et al. [121] proposed a techno-economic analysis of an off-grid photovoltaic with graphite/LiCoO2 storage used to supply an anaerobic digestion biogas power plant (AD). The main contribution is the economical study of the hybrid system including the battery degradation costs. An optimal operating regime is developed for the hybrid system, followed by a study on the levelized cost of electricity (LCOE).

Figure 3 presents a summary of the energy management methodologies used for the MGs based on the above-reviewed literature. Different researches have proposed several methodologies related to energy management in MGs. Many methods are based on classical approaches such as mixed integer linear and nonlinear programming. Linear programming can be considered a good approach depending on objective and constraints, while artificial intelligence methods are focused to approach situations where other methods lead to unsatisfactory results, including renewable generation forecasting and optimal operation of energy storage considering battery aging, among others.

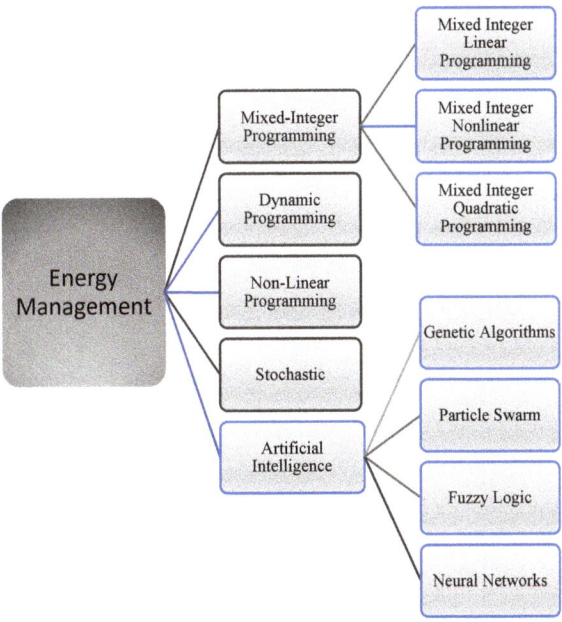

Figure 3. Energy management methodologies in microgrids (MGs) [25].

3.9. Optimization Techniques

Different optimization techniques are generally applied to maximize the power output of each particular source, minimize electricity costs, or maximize storage systems. Figure 4 presents the most commonly employed optimization techniques and algorithms presented in the literature review. Main advantages and disadvantages are briefly presented in Table 2.

Various techniques have been used by different researches. Energy management and the optimization of control in a MG can have one or more objective functions. These functions can vary depending on the optimization problem presented. This can result in a mono-objective or multi-objective problem, which can include the minimization of costs (operation and maintenance cost, fuel cost, and degradation cost of storage elements such as batteries or capacitors), minimization of the emissions and minimization of the unmet load. Table 3 shows a comparison between the different optimization and management methods used in the MGs. Different researchers have proposed metaheuristic techniques to solve the problem of optimization due to multi-constraints, multi-dimensional, and highly nonlinear combinatorial problems. Other authors presented stochastic dynamic programming methods for optimizing the energy management problem with multidimensional objectives. Game theory has been proposed for some researchers to solve problems with conflicting objective functions.

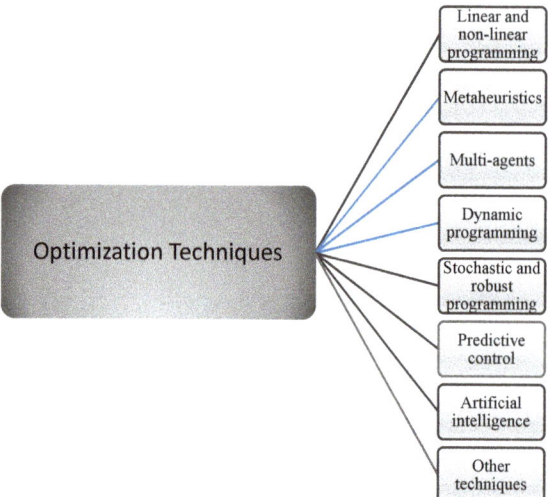

Figure 4. Optimization techniques in microgrid energy management.

Table 2. Comparative analysis of optimization mathematical models.

Optimization Mathematical Model	Advantages	Disadvantages
MILP	Linear programming (LP) is a fast way to solve the problems and the linear constraints result in a convex feasible region, being guaranteed in many cases to obtain the global optimum solution.	Reliability and economic stochastical analysis. Limited capabilities for applications with not differentiable and/or continuous objective functions.
MINLP	It uses simple operations to solve complex problems. It can obtain more than one optimal solution to choose from, which is an advantage over the MILP formulation.	High number of iterations (high computational effort).
Dynamic programming (DP)	It can split the problem into subproblems, optimizing each subproblem and therefore solving sequential problems.	Complex implementation due to high number of recursive functions.
Genetic algorithms (GA)	Population-based evolutionary algorithms that include operations such as crossover, mutation, and selection to find the optimal solution. Adequate convergence speed. Widely used in many fields.	Crossover and mutation parameters, and population and stopping criterial parameters must be set.
Particle swarm optimization (PSO)	Good performance in scattering and optimization problems.	High computational complexities.
Artificial bee colony	Robust population-based algorithm simple to implementate. Adequate convergence speed.	Complex formulation.
Artificial Fish Swarm	Few parameters, fast convergence, high accuracy, and flexibility.	Same advantages of GA but without its disadvantages (crossover and mutation).
Bacterial foraging algorithm	Size and non-linearity of the problem does not affect much. Converge to the optimal solution where analytical methods do not converge.	Large and complex search space.

Table 3. Analysis of microgrid optimization techniques.

Reference	Optimization Technique	Contributions	Constraints	Drawbacks	Single/Multi-objective
[31]	No linear and mixed integer programing	Robust optimal EMS MPC-based to obtain the optimal power scheduling for the different generators, including deferrable and dump loads.	Power balance Battery Diesel Generator Renewable Sources Load	Demand and power losses are not considered.	Multi-objective
[32]	Linear and mixed integer linear programming	Energy management strategy based on the combination of three operating strategies (continuous run mode, power sharing mode, and ON/OFF mode).	Battery Generation dispatch	Battery degradation costs in the optimization models are not considered.	Multi-objective
[35]	Mixed integer no linear programming	Reduced the overall operational costs while maintaining a secured operation of the stand-alone MG.	AC power DC power Converter power Load Distributed generators power	Battery storage systems are not considered. Emission cost of distributed generation based on biomass is not considered.	Mono-objective
[45]	Linear programming	Integration of linear programming-based with artificial intelligence techniques to sole multi-objective optimization.	Power Balance Generation limits of distributed generation	High computational complexity. Battery degradation cost is not considered.	Multi-objective
[46]	Particle swarm algorithm (PSO)	Combination of two optimal storage energy units. Less computation time than GA.	Power of the generators Power exchange with the grid Charge/Discharge of the storage units Supply and demand balance	Emission cost of the conventional generator is not considered.	Multi-objective
[48]	Particle swarm algorithm (PSO) with Gaussian mutation	PSO variant new algorithm.	Active power Voltage Current	Power losses are not considered. Emissions of distributed generation are not considered.	Mono-objective
[50]	Artificial bee colony	Two layer control model used to minimize operational cost of a microgrid.	Power balance Dispatchable resources Non-dispatchable resources Storage elements	Complex formulation. Emission cost of a dispatchable microturbine is not considered.	Mono-objective
[51]	Fuzzy logic (Grey Wolf Optimization)	Optimization of the size of the battery energy storage and of the generation plan.	Power balance Power of the generators Battery load	Battery degradation cost is not considered.	Multi-objective
[53]	Evolutionary algorithm (EA) and the particle swarm optimization (PSO) Algorithm	Application of an energy hub model for optimization of a multicarrier MG.	Power balance Voltage in the transformers	Deterministic conditions assumed are a limitation.	Mono-objective

Table 3. *Cont*.

Reference	Optimization Technique	Contributions	Constraints	Drawbacks	Single/Multi-objective
[55]	Artificial fish swarm optimization	An energy management planning of a MG including storage for a whole day is optimized, considering dynamic pricing and demand side management.	Power balance Conventional power generation Conventional power generators Energy storage Utility grid power	Battery degradation cost is not considered.	Mono-objective
[57]	Particle swarm algorithm (PSO)	Three different objectives are considered: Reliability, cost of operation, and environmental impact.	Not specified	Battery degradation cost is not considered.	Multi-objective
[58]	Bacterial foraging algorithm	Optimized the exchanging power with the grid and the generators and battery setpoints. Fast convergence.	Power balance Generation limits of distributed generators Storage limits	Power loss not considered.	Multi-objective
[60]	Mixed-integer nonlinear programming (MINLP)	Reduced dependency on forecast information. Different battery models compared.	Charge flow Dispatch of generators Generator on/off programming Charge/Discharge of batteries	Battery lifetime prediction is ignored.	Multi-objective
[61]	Dynamic Rules	MG management system uses different limits for the SOC of the batteries bank.	Battery Power balance	Battery cost and degradation are not considered.	Mono-objective
[64]	Dynamic programming	Energy management strategy for PV. Batteries to stabilize and permit PV to run at a constant and stable output power.	Charge/Discharge of batteries	Battery degradation and lifetime prediction are not considered.	Multi-objective
[70]	Multi-agents	Efficient strategy for real-time management of energy storage used to compensate power mismatch optimally.	Charge/Discharge of batteries Load Scheduling Power Balance	Prediction of battery ageing is not included.	Multi-objective
[72]	Multi-agents	Control scheme composed of several levels with coordinated control.	Charge/Discharge of batteries	High complexity control scheme.	Multi-objective
[73]	Multi-agents	Battery energy storage system, optimization problem based on distributed intelligence, and a multi-agent system.	Not specified	Battery degradation is not considered.	Multi-objective
[80]	Mixed integer programming	Dual-stage optimization. First stage determines hourly unit commitment of the generators, the second stage performs economic dispatch of the generators and batteries.	Startup costs of renewable energy and conventional generators	Power losses are not considered. Battery degradation is not considered.	Mono-objective

Table 3. *Cont.*

Reference	Optimization Technique	Contributions	Constraints	Drawbacks	Single/Multi-objective
[84]	Stochastic	A simple method to incorporate the impact of the scheduling in stand-alone mode on the grid-connected operation.	Power balance Dispatchable Distributed generation Renewable power generation Load Charge/Discharge of batteries	Battery ageing model is not considered. Emission cost of DG are not taken into consideration.	Mono-objective
[89]	Robust programming	Optimization of load management for hybrid Wind–Battery–Diesel systems.	Power Diesel Generator Power wind turbine Power battery bank	Controllable loads shifting can be non-optimal.	Mono-objective
[91]	Mixed Integer Quadratic Programming	Integrated stochastic energy management model, simultaneously considering unit commitment for generators and demand side management.	Power balance Generation Demand Reserve capacity	Computational time is higher than that of deterministic model. Emission cost of conventional generators and DG are not taken into consideration.	Mono-objective
[92]	Model predictive control	Automated load shedding of noncritical loads when foreseeable power unbalances could affect the stability of the MG.	Power distributed generators	The charging and discharging rates of the batteries were not considered. It does not consider communication delays.	Multi-objective
[97]	Model predictive control	A comprehensive mathematical formulation of the optimal EMS for stand-alone microgrids, considering power flow and unit commitment operational constraints.	Power balance Reserve Unit commitment Energy storage Grid DG	Higher computational burden and complexity. Emission cost of conventional generators is not considered.	Mono-objective
[101]	Model predictive control	The main contribution of this work is daily optimization that accounts for the degradation of the lead-acid battery by corrosion and capacity losses.	Not specified	Lithium battery model is not considered.	Multi-objective
[103]	Genetic algorithm	A novel cost function is including costs of selling and buying power, and the start-up costs of distributed resources.	Power balance Emissions Battery storage Startup and downtime of generators	Distributed sources and battery state of charge are not considered. The uncertainty in energy generation by the MGs and the uncertainty in customers are not considered.	Mono-objective
[104]	Game theory	In multiple MGs, distributed energy management schedule.	Energy exchange with the grid Generation capacity of the MG	Computational complexity is not discussed.	Multi-objective

Table 3. *Cont.*

Reference	Optimization Technique	Contributions	Constraints	Drawbacks	Single/Multi-objective
[111]	Artificial Intelligence (Fuzzy logic)	Simple implementation, improved the grid power profile quality.	Charge/Discharge of batteries	Only the battery charger/grid-connected inverter is controllable. Battery degradation is not considered.	Multi-objective
[114]	Game theory	Minimize fuel cost and trading power cost.	Power balance DG Conventional generator power Limit for the transferable power between The main grid and MG	Emission cost of conventional generators is not considered.	Multi-objective
[118]	Markov decision process	Linear model to evaluate the MG lifetime cost.	Gas turbine capacity Gas turbine emissions	Number of possible combinations of sizes is limited.	Mono-objective
[121]	Rule-based	Study on the economic projection of the hybrid system with the battery degradation costs. An optimal operating regime is developed for hybrid system, followed by a study on the levelized cost of electricity (LCOE). Accuracy of degradation costs of the energy storage.	Power balance SOC battery	Temperature not considered in the capacity fade model. Dynamic state of charge cycling conditions not considered.	Multi-objective

3.10. Microgrid Operating Modes

A considerable number of papers have been published on interconnected microgrids, while discussing various modes of microgrid operation. On the other hand, the stand-alone mode is considered by many authors as an alternative supply measure mainly in the rural areas or regions with no conventional grids [122]. Thus, both the on- and off-grid operating modes are a feasible alternative. Table 4 summarizes the above considerations.

Table 4. Microgrid operating modes.

Reference	Microgrid Mode Operation
[11,20,30,32,33,36–39,45,49,51–53,55,56,58,59,63,67–71, 74,77–81,83,85–87,91,93,94,96,99,100,103–114,117,118]	Grid-Connected
[9,31,34,40,42,44,47,48,50,54,57,60,62,65,66,72,73,75,76, 82,84,88–90,92,97,98,101,102,115,116,119,121]	Off-Grid
[8,15,19,35,43,46,61,64,95,120,122]	Grid-Connected/Off-Grid

3.11. Modelling and Simulation Tools

Table 5 presents a summary of the most popular simulation tools, wherein tools such as Matlab/Simulink (MathWorks, Natick, MA, USA) and MATPOWER have particular importance. Matlab is a numerical computing environment of 4th generation programming language, it can interface with other languages such as C, C++, C#, Java, Fortran, and Python. MATPOWER is an open-source tool that is used to simulate optimal power flows, which uses Monte Carlo to evaluate the performance of MG. Alternately, other tools such as GAMS, which is an optimization language for linear, nonlinear, and mixed programming, have been used by many authors to solve the uncertainty problem in energy management and for optimal dimensioning of the microgrid. Other tools such as CPLEX have been employed, which is an optimizer based on the C language and is compatible with other languages like C++, Java, and Python.

Table 5. Simulation software and tools used in the management of microgrids.

References	Tools	Characteristics
[61]	PSCAD/EMTDC	Simulation software power systems, power electronics, HVDC, FACTS, and control system.
[11,32,33,35,38,62,64,65, 67,70,77,93,97,104,109, 110,121]	MATLAB/Simulink MATPOWER	Matrix based programming language used by engineers in power systems, power electronics, telecommunications, and control, among others. Compatible with other programming languages (C++, Java, and fortran).
[30,76]	GAMS (GAMS Development Corp., Fairfax, VA, USA)	High level language for mathematical optimization of mixed integer linear and nonlinear.
[74]	C++	Application development environment of C++ for Windows.
[40]	TRNSYS (Thermal Energy System Specialists, LLC, Madison, WI, USA) HOMER HOGA	Simulation software to model hybrid systems of energy generation. Hybrid Optimization by Genetic Algorithms.
[75]	RSCAD (RTDS Technologies Inc., Winnipeg, MA, Canada) JADE (Jade, Christchurch, New Zealand)	Real time simulator for power systems.
[67,68,71,72]	JADE	Java environment platform for multi-agents.
[30,118,122]	HOMER	Simulation software to model hybrid systems of energy generation.
[36,83,103]	CPLEX (IBM, Armonk, NY, USA)	Optimization software compatible with C, C++, Java, and Python languages.

The simulation and modeling of microgrids has been analyzed with programs such as Simulink and PSCAD/EMTDC (Manitoba Hydro International Ltd., Winnipeg, Manitoba, Canada). Both tools are used for power control and energy management in microgrids.

Software such as HOMER (Pro Version, HOMER Energy LLC, Boulder, CO, USA), HOGA (or its updated version iHOGA) (Pro+ Version, University of Zaragoza, Zaragoza, Spain), or HYBRID2 (University of Massachusets, NREL/NWTC, Golden, CO, USA) also deserve a mention, which can be used to optimize the operation and energy management of hybrid systems with renewable energies.

4. Conclusions and Future Research

The literature review highlighted two approaches for microgrid energy management: The centralized and decentralized approaches. The first incorporates optimization using the available information in the absence of a coordination strategy between the actors in a microgrid. A computer centre transmits the optimal settings to each participant. The second approach implements optimization using partial information and a strategy for coordinating the microgrid participants; each participant evaluates its own optimal settings. Centralized management is mostly implemented in metaheuristic methods, and decentralized management is frequently implemented in methods based on multi-agents. Many publications have proposed centralized management for microgrids. However, the incursion of distributed energy resources (DER) may cause this type of management to face issues when implemented in a centralized information system because there might be a demand for high computational cost due to the large quantity of data. Distributed energy management may be an alternative solution to this problem. It solves the problem of data processing and reduces processing needs by using distributed controllers that manage the data in real time and require communication equipment that might result in additional costs (for e.g., Bluetooth, Wi-Fi, wireless networks, and IoT).

An energy management model for a microgrid includes data acquisition systems, supervised control, human machine interface (HMI), and the monitoring and data analysis of meteorological variables.

The literature review mainly presented management methods based on foresight and short-term management. The choice of centralized or decentralized management ensures that the microgrid designer and operator realize a cost–benefit balance. This enables one to determine the management model that is most convenient for the microgrid. Though decentralized management offers more flexibility, an integral analysis is necessary to ensure reliable and safe system operation.

The energy management problem or optimization control in a microgrid becomes a mono-objective management/optimization model when a single cost function is presented. This function typically corresponds to the operating cost of the microgrids. The problem becomes a multi-objective management/optimization model when it simultaneously presents a solution to the technical, economic, and environmental problems. Based on the literature, different authors have addressed the problem and provided solutions using methods such as the classic ones with linear and nonlinear programming, heuristic methods, predictive control, dynamic programming, agent-based methods, and artificial intelligence. These methods are chosen based on their practicality, reliability, and resource availability in the microgrid environment.

With regard to storage systems in microgrids, lithium batteries can be an important alternative to lead-acid batteries in the future. The advantages of Li-ion batteries compared to lead-acid batteries are a long cycle life, fast charging, high energy density, and low maintenance. Currently, lead acid batteries are economically better than Li-ion batteries when used in microgrids, but a decrease in the acquisition cost of lithium batteries is expected in the coming years that will cause them to be competitive with those of lead-acid. Thus, further research on the optimal energy management of energy systems and the management of lithium batteries is required while considering more accurate degradation models to accurately predict the battery lifetime in real operating conditions.

Author Contributions: Conceptualization, Y.E.G.V. and R.D.-L.; methodology, Y.E.G.V. and R.D.-L.; formal analysis, J.L.B.-A.; investigation, Y.E.G.V.; resources, Y.E.G.V. and R.D.-L.; data curation, R.D.-L.; writing—original draft preparation, Y.E.G.V.; writing—review and editing, R.D.-L. and J.L.B.-A.; visualization, Y.E.G.V. and R.D.-L.; supervision, R.D.-L. and J.L.B.-A.; project administration, R.D.-L.; funding acquisition, R.D.-L. and J.L.B.-A.

Funding: This research was funded by Government of Aragon "Gobierno de Aragón. Grupo de referencia Gestión estratégica de la energía eléctrica", grant number 28850.

Conflicts of Interest: The authors declare no conflict of interest.

Abbreviations

MG	Microgrid
AC	Alternating current line
ARMA	Auto-regressive moving average models
CSA	Crow search algorithm
DC	Direct current line
DG	Distributed generation
DER	Distributed energy resources
EEMS	Expert system for energy management
EMS	Energy management system
GAMS	General algebraic modeling system
HMI	Human machine interfaces
HOGA	Hybrid optimization by genetic algorithms
HOMER	Hybrid optimization model for multiple energy resources
iHOGA	Improved Hybrid optimization by genetic algorithms
JADE	Java platform for agent developers
MGSC	Microgrid supervisory controllers
MILP	Mixed integer linear programming
MO	Multi-objective
MPC	Model predictive control
PSO	Particle swarm optimization
PV	Photovoltaic
VPP	Virtual power plant

References

1. Wu, J.; Yan, J.; Jia, H.; Hatziargyriou, N.; Djilali, N.; Sun, H. Integrated Energy Systems. *Appl. Energy* **2016**, *167*, 155–157. [CrossRef]
2. Renewables. Int Energy Agency, IEA. 2019. Available online: https://www.iea.org/topics/renewables/ (accessed on June 2019).
3. Parhizi, S.; Lotfi, H.; Khodaei, A.; Bahramirad, S. State of the art in research on microgrids: A review. *IEEE Access* 2015. [CrossRef]
4. Caspary, G. Gauging the future competitiveness of renewable energy in Colombia. *Energy Econ.* **2009**, *31*, 443–449. [CrossRef]
5. Afgan, N.H.; Carvalho, M.G. Sustainability assessment of a hybrid energy system. *Energy Policy* **2008**, *36*, 2903–2910. [CrossRef]
6. Faccio, M.; Gamberi, M.; Bortolini, M.; Nedaei, M. State-of-art review of the optimization methods to design the configuration of hybrid renewable energy systems (HRESs). *Front. Energy* **2018**, *12*, 591–622. [CrossRef]
7. Nema, P.; Nema, R.K.; Rangnekar, S. A current and future state of art development of hybrid energy system using wind and PV-solar: A review. *Renew. Sustain. Energy Rev.* **2009**. [CrossRef]
8. Lujano Rojas, J.M. Análisis y gestión óptima de la demanda en sistemas eléctricos conectados a la red y en sistemas aislados basados en fuentes renovables. Ph.D. Thesis, Univesity of Zaragoza, Zaragoza, Spain, 2012.
9. Cristóbal-Monreal, I.R.; Dufo-López, R. Optimisation of photovoltaic-diesel-battery stand-alone systems minimising system weight. *Energy Convers. Manag.* **2016**. [CrossRef]
10. Lasseter, R.H. MicroGrids. In Proceedings of the 2002 IEEE Power Engineering Society Winter Meeting, New York, NY, USA, 27–31 January 2002; pp. 305–308. [CrossRef]

11. Thirugnanam, K.; Kerk, S.K.; Yuen, C.; Liu, N.; Zhang, M. Energy Management for Renewable Microgrid in Reducing Diesel Generators Usage with Multiple Types of Battery. *IEEE Trans. Ind. Electron.* **2018**. [CrossRef]
12. Yang, N.; Paire, D.; Gao, F.; Miraoui, A. Power management strategies for microgrid—A short review. In Proceedings of the 2013 IEEE Industry Applications Society Annual Meeting, Lake Buena Vista, FL, USA, 6–11 October 2013. [CrossRef]
13. Atcitty, S.; Neely, J.; Ingersoll, D.; Akhil, A.; Waldrip, K. Battery Energy Storage System. *Green Energy Technol.* **2013**. [CrossRef]
14. Lasseter, R.H. CERTS Microgrid. In Proceedings of the 2007 IEEE International Conference on System of Systems Engineering, San Antonio, TX, USA, 16–18 April 2007. [CrossRef]
15. Hatziargyriou, N.; Asano, H.; Iravani, R.; Marnay, C. Microgrids: An Overview of Ongoing Research, Development, and Demonstration Projects. *IEEE Power Energy Mag.* **2007**. [CrossRef]
16. Shi, W.; Lee, E.K.; Yao, D.; Huang, R.; Chu, C.C.; Gadh, R. Evaluating microgrid management and control with an implementable energy management system. In Proceedings of the 2014 IEEE International Conference on Smart Grid Communications (SmartGridComm), Venice, Italy, 15 January 2015. [CrossRef]
17. Shi, W.; Li, N.; Chu, C.C.; Gadh, R. Real-Time Energy Management in Microgrids. *IEEE Trans. Smart Grid* **2017**, *8*, 228–238. [CrossRef]
18. Stanton, K.N.; Giri, J.C.; Bose, A. Energy management. *Syst. Control Embed. Syst. Energy Mach.* **2017**. [CrossRef]
19. Su, W.; Wang, J. Energy Management Systems in Microgrid Operations. *Electr. J.* **2012**. [CrossRef]
20. Gildardo Gómez, W.D. Metodología para la Gestión Óptima de Energía en una Micro red Eléctrica Interconectada. Ph.D. Thesis, Universidad Nacional de Colombia, Medellín, Colombia, 2016.
21. Zia, M.F.; Elbouchikhi, E.; Benbouzid, M. Microgrids energy management systems: A critical review on methods, solutions, and prospects. *Appl. Energy* **2018**. [CrossRef]
22. Robert, F.C.; Sisodia, G.S.; Gopalan, S. A critical review on the utilization of storage and demand response for the implementation of renewable energy microgrids. *Sustain. Cities Soc.* **2018**. [CrossRef]
23. Olatomiwa, L.; Mekhilef, S.; Ismail, M.S.; Moghavvemi, M. Energy management strategies in hybrid renewable energy systems: A review. *Renew. Sustain. Energy Rev.* **2016**. [CrossRef]
24. Meng, L.; Sanseverino, E.R.; Luna, A.; Dragicevic, T.; Vasquez, J.C.; Guerrero, J.M. Microgrid supervisory controllers and energy management systems: A literature review. *Renew. Sustain. Energy Rev.* **2016**. [CrossRef]
25. Ahmad Khan, A.; Naeem, M.; Iqbal, M.; Qaisar, S.; Anpalagan, A. A compendium of optimization objectives, constraints, tools and algorithms for energy management in microgrids. *Renew. Sustain. Energy Rev.* **2016**. [CrossRef]
26. Gamarra, C.; Guerrero, J.M. Computational optimization techniques applied to microgrids planning: A review. *Renew. Sustain. Energy Rev.* **2015**. [CrossRef]
27. Fathima, A.H.; Palanisamy, K. Optimization in microgrids with hybrid energy systems—A review. *Renew. Sustain. Energy Rev.* **2015**. [CrossRef]
28. Suchetha, C.; Ramprabhakar, J. Optimization techniques for operation and control of microgrids—Review. *J. Green Eng.* **2018**. [CrossRef]
29. Lee, E.K.; Shi, W.; Gadh, R.; Kim, W. Design and implementation of a microgrid energy management system. *Sustainability* **2016**, *8*, 1143. [CrossRef]
30. Ahmad, J.; Imran, M.; Khalid, A.; Iqbal, W.; Ashraf, S.R.; Adnan, M.; Ali, S.F.; Khokhar, K.S. Techno economic analysis of a wind-photovoltaic-biomass hybrid renewable energy system for rural electrification: A case study of Kallar Kahar. *Energy* **2018**. [CrossRef]
31. Taha, M.S.; Mohamed, Y.A.R.I. Robust MPC-based energy management system of a hybrid energy source for remote communities. In Proceedings of the 2016 IEEE Electrical Power and Energy Conference (EPEC), Ottawa, ON, Canada, 12–14 October 2016. [CrossRef]
32. Sukumar, S.; Mokhlis, H.; Mekhilef, S.; Naidu, K.; Karimi, M. Mix-mode energy management strategy and battery sizing for economic operation of grid-tied microgrid. *Energy* **2017**. [CrossRef]
33. Paul, T.G.; Hossain, S.J.; Ghosh, S.; Mandal, P.; Kamalasadan, S. A Quadratic Programming Based Optimal Power and Battery Dispatch for Grid-Connected Microgrid. *IEEE Trans. Ind. Appl.* **2018**. [CrossRef]

34. Delgado, C.; Dominguez-Navarro, J.A. Optimal design of a hybrid renewable energy system. In Proceedings of the 2014 Ninth International Conference on Ecological Vehicles and Renewable Energies (EVER), Monte-Carlo, Monaco, 25–27 March 2014. [CrossRef]
35. Helal, S.A.; Najee, R.J.; Hanna, M.O.; Shaaban, M.F.; Osman, A.H.; Hassan, M.S. An energy management system for hybrid microgrids in remote communities. *Can. Conf. Electr. Comput. Eng.* **2017**. [CrossRef]
36. Umeozor, E.C.; Trifkovic, M. Energy management of a microgrid via parametric programming. *IFAC-PapersOnLine* **2016**. [CrossRef]
37. Xing, X.; Meng, H.; Xie, L.; Li, P.; Toledo, S.; Zhang, Y.; Guerrero, J.M. Multi-time-scales energy management for grid-on multi-layer microgrids cluster. In Proceedings of the 2017 IEEE Southern Power Electronics Conference (SPEC), Puerto Varas, Chile, 9 April 2018. [CrossRef]
38. Correa, C.A.; Marulanda, G.; Garces, A. Optimal microgrid management in the Colombian energy market with demand response and energy storage. In Proceedings of the 2016 IEEE Power and Energy Society General Meeting (PESGM), Boston, MA, USA, 14 November 2016. [CrossRef]
39. Cardoso, G.; Brouhard, T.; DeForest, N.; Wang, D.; Heleno, M.; Kotzur, L. Battery aging in multi-energy microgrid design using mixed integer linear programming. *Appl. Energy* **2018**. [CrossRef]
40. Behzadi, M.S.; Niasati, M. Comparative performance analysis of a hybrid PV/FC/battery stand-alone system using different power management strategies and sizing approaches. *Int. J. Hydrogen Energy* **2015**. [CrossRef]
41. Dufo-López, R.; Bernal-Agustín, J.L.; Contreras, J. Optimization of control strategies for stand-alone renewable energy systems with hydrogen storage. *Renew Energy* **2007**. [CrossRef]
42. Das, B.K.; Al-Abdeli, Y.M.; Kothapalli, G. Effect of load following strategies, hardware, and thermal load distribution on stand-alone hybrid CCHP systems. *Appl. Energy* **2018**. [CrossRef]
43. Luna, A.C.; Meng, L.; Diaz, N.L.; Graells, M.; Vasquez, J.C.; Guerrero, J.M. Online Energy Management Systems for Microgrids: Experimental Validation and Assessment Framework. *IEEE Trans. Power Electron.* **2018**. [CrossRef]
44. Chalise, S.; Sternhagen, J.; Hansen, T.M.; Tonkoski, R. Energy management of remote microgrids considering battery lifetime. *Electr. J.* **2016**. [CrossRef]
45. Chaouachi, A.; Kamel, R.M.; Andoulsi, R.; Nagasaka, K. Multiobjective intelligent energy management for a microgrid. *IEEE Trans. Ind. Electron.* **2013**, *60*, 1688–1699. [CrossRef]
46. Li, H.; Eseye, A.T.; Zhang, J.; Zheng, D. Optimal energy management for industrial microgrids with high-penetration renewables. *Prot. Control Mod. Power Syst.* **2017**. [CrossRef]
47. Nivedha, R.R.; Singh, J.G.; Ongsakul, W. PSO based economic dispatch of a hybrid microgrid system. In Proceedings of the 4th 2018 International Conference on Power, Signals, Control and Computation (EPSCICON 2018), Thrissur, India, 6–10 January 2018. [CrossRef]
48. Abedini, M.; Moradi, M.H.; Hosseinian, S.M. Optimal management of microgrids including renewable energy scources using GPSO-GM algorithm. *Renew. Energy* **2016**. [CrossRef]
49. Nikmehr, N.; Najafi-Ravadanegh, S. Optimal operation of distributed generations in micro-grids under uncertainties in load and renewable power generation using heuristic algorithm. *IET Renew. Power Gener.* **2015**. [CrossRef]
50. Marzband, M.; Azarinejadian, F.; Savaghebi, M.; Guerrero, J.M. An optimal energy management system for islanded microgrids based on multiperiod artificial bee colony combined with markov chain. *IEEE Syst. J.* **2017**. [CrossRef]
51. Ei-Bidairi, K.S.; Nguyen, H.D.; Jayasinghe, S.D.G.; Mahmoud, T.S. Multiobjective Intelligent Energy Management Optimization for Grid-Connected Microgrids. In Proceedings of the 2018 IEEE International Conference on Environment and Electrical Engineering and 2018 IEEE Industrial and Commercial Power Systems Europe (EEEIC / I&CPS Europe), Palermo, Italy, 12–15 June 2018. [CrossRef]
52. Papari, B.; Edrington, C.S.; Vu, T.V.; Diaz-Franco, F. A heuristic method for optimal energy management of DC microgrid. In Proceedings of the 2017 IEEE Second International Conference on DC Microgrids (ICDCM), Nuremburg, Germany, 27–29 June 2017. [CrossRef]
53. Wasilewski, J. Optimisation of multicarrier microgrid layout using selected metaheuristics. *Int. J. Electr. Power Energy Syst.* **2018**. [CrossRef]
54. Ogunjuyigbe, A.S.O.; Ayodele, T.R.; Akinola, O.A. Optimal allocation and sizing of PV/Wind/Split-diesel/Battery hybrid energy system for minimizing life cycle cost, carbon emission and dump energy of remote residential building. *Appl. Energy* **2016**. [CrossRef]

55. Kumar, K.P.; Saravanan, B. Day ahead scheduling of generation and storage in a microgrid considering demand Side management. *J. Energy Storage* **2019**. [CrossRef]
56. Hossain, M.A.; Pota, H.R.; Squartini, S.; Abdou, A.F. Modified PSO algorithm for real-time energy management in grid-connected microgrids. *Renew. Energy* **2019**. [CrossRef]
57. Azaza, M.; Wallin, F. Multi objective particle swarm optimization of hybrid micro-grid system: A case study in Sweden. *Energy* **2017**. [CrossRef]
58. Motevasel, M.; Seifi, A.R. Expert energy management of a micro-grid considering wind energy uncertainty. *Energy Convers. Manag.* **2014**. [CrossRef]
59. Rouholamini, M.; Mohammadian, M. Heuristic-based power management of a grid-connected hybrid energy system combined with hydrogen storage. *Renew. Energy* **2016**. [CrossRef]
60. Shuai, H.; Fang, J.; Ai, X.; Wen, J.; He, H. Optimal Real-Time Operation Strategy for Microgrid: An ADP-Based Stochastic Nonlinear Optimization Approach. *IEEE Trans. Sustain. Energy* **2019**. [CrossRef]
61. Almada, J.B.; Leão, R.P.S.; Sampaio, R.F.; Barroso, G.C. A centralized and heuristic approach for energy management of an AC microgrid. *Renew. Sustain. Energy Rev.* **2016**. [CrossRef]
62. Wu, N.; Wang, H. Deep learning adaptive dynamic programming for real time energy management and control strategy of micro-grid. *J. Clean Prod.* **2018**. [CrossRef]
63. Zhuo, W. Microgrid energy management strategy with battery energy storage system and approximate dynamic programming. In Proceedings of the 2018 37th Chinese Control Conference (CCC), Wuhan, China, 25–27 July 2018. [CrossRef]
64. Choudar, A.; Boukhetala, D.; Barkat, S.; Brucker, J.M. A local energy management of a hybrid PV-storage based distributed generation for microgrids. *Energy Convers Manag.* **2015**. [CrossRef]
65. Merabet, A.; Tawfique Ahmed, K.; Ibrahim, H.; Beguenane, R.; Ghias, A.M.Y.M. Energy Management and Control System for Laboratory Scale Microgrid Based Wind-PV-Battery. *IEEE Trans. Sustain. Energy* **2017**. [CrossRef]
66. Luu, N.A.; Tran, Q.T.; Bacha, S. Optimal energy management for an island microgrid by using dynamic programming method. In Proceedings of the 2015 IEEE Eindhoven PowerTech, Eindhoven, The Netherlands, 29 June–2 July 2015. [CrossRef]
67. Boudoudouh, S.; Maâroufi, M. Multi agent system solution to microgrid implementation. *Sustain. Cities Soc.* **2018**. [CrossRef]
68. Raju, L.; Morais, A.A.; Rathnakumar, R.; Ponnivalavan, S.; Thavam, L.D. Micro-grid grid outage management using multi-agent systems. In Proceedings of the 2017 Second International Conference on Recent Trends and Challenges in Computational Models (ICRTCCM), Tindivanam, India, 3–4 February 2017. [CrossRef]
69. Bogaraj, T.; Kanakaraj, J. Intelligent energy management control for independent microgrid. *Sadhana—Acad. Proc. Eng. Sci.* **2016**. [CrossRef]
70. Anvari-Moghaddam, A.; Rahimi-Kian, A.; Mirian, M.S.; Guerrero, J.M. A multi-agent based energy management solution for integrated buildings and microgrid system. *Appl. Energy* **2017**. [CrossRef]
71. Kumar Nunna, H.S.V.S.; Doolla, S. Energy management in microgrids using demand response and distributed storage—A multiagent approach. *IEEE Trans. Power. Deliv.* **2013**. [CrossRef]
72. Dou, C.X.; Liu, B. Multi-agent based hierarchical hybrid control for smart microgrid. *IEEE Trans. Smart Grid* **2013**. [CrossRef]
73. Karavas, C.S.; Kyriakarakos, G.; Arvanitis, K.G.; Papadakis, G. A multi-agent decentralized energy management system based on distributed intelligence for the design and control of autonomous polygeneration microgrids. *Energy Convers. Manag.* **2015**. [CrossRef]
74. Mao, M.; Jin, P.; Hatziargyriou, N.D.; Chang, L. Multiagent-based hybrid energy management system for microgrids. *IEEE Trans. Sustain. Energy* **2014**. [CrossRef]
75. Netto, R.S.; Ramalho, G.R.; Bonatto, B.D.; Carpinteiro, O.A.S.; Zambroni De Souza, A.C.; Oliveira, D.Q.; Aparecido da Silva Braga, R. Real-Time framework for energy management system of a smart microgrid using multiagent systems. *Energies* **2018**. [CrossRef]
76. Hu, M.C.; Lu, S.Y.; Chen, Y.H. Stochastic programming and market equilibrium analysis of microgrids energy management systems. *Energy* **2016**. [CrossRef]
77. Reddy, S.S. Optimization of renewable energy resources in hybrid energy systems. *J. Green Eng.* **2017**. [CrossRef]

78. Lu, T.; Ai, Q.; Wang, Z. Interactive game vector: A stochastic operation-based pricing mechanism for smart energy systems with coupled-microgrids. *Appl. Energy* **2018**. [CrossRef]
79. Xiang, Y.; Liu, J.; Liu, Y. Robust Energy Management of Microgrid with Uncertain Renewable Generation and Load. *IEEE Trans. Smart Grid* **2016**. [CrossRef]
80. Hu, W.; Wang, P.; Gooi, H.B. Towards optimal energy management of microgrids with a realistic model. In Proceedings of the 2016 Power Systems Computation Conference (PSCC), Genoa, Italy, 20–24 June 2016. [CrossRef]
81. Shen, J.; Jiang, C.; Liu, Y.; Wang, X. A Microgrid Energy Management System and Risk Management under an Electricity Market Environment. *IEEE Access* **2016**. [CrossRef]
82. Rezaei, N.; Kalantar, M. Stochastic frequency-security constrained energy and reserve management of an inverter interfaced islanded microgrid considering demand response programs. *Int. J. Electr. Power Energy Syst.* **2015**. [CrossRef]
83. Su, W.; Wang, J.; Roh, J. Stochastic energy scheduling in microgrids with intermittent renewable energy resources. *IEEE Trans. Smart Grid* **2014**. [CrossRef]
84. Farzin, H.; Fotuhi-Firuzabad, M.; Moeini-Aghtaie, M. Stochastic Energy Management of Microgrids during Unscheduled Islanding Period. *IEEE Trans. Ind. Inform.* **2017**. [CrossRef]
85. Liu, J.; Chen, H.; Zhang, W.; Yurkovich, B.; Rizzoni, G. Energy Management Problems under Uncertainties for Grid-Connected Microgrids: A Chance Constrained Programming Approach. *IEEE Trans. Smart Grid* **2017**. [CrossRef]
86. Kuznetsova, E.; Li, Y.F.; Ruiz, C.; Zio, E. An integrated framework of agent-based modelling and robust optimization for microgrid energy management. *Appl. Energy* **2014**. [CrossRef]
87. Zachar, M.; Daoutidis, P. Energy management and load shaping for commercial microgrids coupled with flexible building environment control. *J. Energy Storage* **2018**. [CrossRef]
88. Battistelli, C.; Agalgaonkar, Y.P.; Pal, B.C. Probabilistic Dispatch of Remote Hybrid Microgrids Including Battery Storage and Load Management. *IEEE Trans. Smart Grid* **2017**. [CrossRef]
89. Lujano-Rojas, J.M.; Monteiro, C.; Dufo-López, R.; Bernal-Agustín, J.L. Optimum load management strategy for wind/diesel/battery hybrid power systems. *Renew. Energy* **2012**. [CrossRef]
90. Zhai, M.; Liu, Y.; Zhang, T.; Zhang, Y. Robust model predictive control for energy management of isolated microgrids. *IEEE Int. Conf. Ind. Eng. Eng. Manag.* **2018**. [CrossRef]
91. Zhang, Y.; Meng, F.; Wang, R.; Zhu, W.; Zeng, X.J. A stochastic MPC based approach to integrated energy management in microgrids. *Sustain. Cities Soc.* **2018**. [CrossRef]
92. Minchala-Avila, L.I.; Garza-Castanon, L.; Zhang, Y.; Ferrer, H.J.A. Optimal Energy Management for Stable Operation of an Islanded Microgrid. *IEEE Trans. Ind. Inform.* **2016**. [CrossRef]
93. Ju, C.; Wang, P.; Goel, L.; Xu, Y. A two-layer energy management system for microgrids with hybrid energy storage considering degradation costs. *IEEE Trans. Smart Grid* **2018**. [CrossRef]
94. Valencia, F.; Collado, J.; Sáez, D.; Marín, L.G. Robust Energy Management System for a Microgrid Based on a Fuzzy Prediction Interval Model. *IEEE Trans. Smart Grid* **2016**. [CrossRef]
95. Ganesan, S.; Padmanaban, S.; Varadarajan, R.; Subramaniam, U.; Mihet-Popa, L. Study and analysis of an intelligent microgrid energy management solution with distributed energy sources. *Energies* **2017**. [CrossRef]
96. Garcia-Torres, F.; Bordons, C. Optimal Economical Schedule of Hydrogen-Based Microgrids With Hybrid Storage Using Model Predictive Control. *IEEE Trans. Ind. Electron.* **2015**. [CrossRef]
97. Solanki, B.V.; Raghurajan, A.; Bhattacharya, K.; Canizares, C.A. Including Smart Loads for Optimal Demand Response in Integrated Energy Management Systems for Isolated Microgrids. *IEEE Trans. Smart Grid* **2017**. [CrossRef]
98. Oh, S.; Chae, S.; Neely, J.; Baek, J.; Cook, M. Efficient model predictive control strategies for resource management in an islanded microgrid. *Energies* **2017**. [CrossRef]
99. Prodan, I.; Zio, E.; Stoican, F. Fault tolerant predictive control design for reliable microgrid energy management under uncertainties. *Energy* **2015**. [CrossRef]
100. Wu, Z.; Tazvinga, H.; Xia, X. Demand side management of photovoltaic-battery hybrid system. *Appl. Energy* **2015**. [CrossRef]

101. Dufo-López, R.; Fernández-Jiménez, L.A.; Ramírez-Rosado, I.J.; Artal-Sevil, J.S.; Domínguez-Navarro, J.A.; Bernal-Agustín, J.L. Daily operation optimisation of hybrid stand-alone system by model predictive control considering ageing model. *Energy Convers. Manag.* **2017**. [CrossRef]
102. Schiffer, J.; Sauer, D.U.; Bindner, H.; Cronin, T.; Lundsager, P.; Kaiser, R. Model prediction for ranking lead-acid batteries according to expected lifetime in renewable energy systems and autonomous power-supply systems. *J. Power Sources* **2007**. [CrossRef]
103. Elsied, M.; Oukaour, A.; Gualous, H.; Hassan, R. Energy management and optimization in microgrid system based ongreen energy. *Energy* **2015**. [CrossRef]
104. Mondal, A.; Misra, S.; Patel, L.S.; Pal, S.K.; Obaidat, M.S. DEMANDS: Distributed energy management using noncooperative scheduling in smart grid. *IEEE Syst. J.* **2018**. [CrossRef]
105. Prathyush, M.; Jasmin, E.A. Fuzzy Logic Based Energy Management System Design for AC Microgrid. In Proceedings of the International Conference on Inventive Communication and Computational Technologies (ICICCT), Coimbatore, India, 20 April 2018. [CrossRef]
106. Leonori, S.; Rizzi, A.; Paschero, M.; Mascioli, F.M.F. Microgrid Energy Management by ANFIS Supported by an ESN Based Prediction Algorithm. In Proceedings of the International Joint Conference on Neural Networks (IJCNN), Rio de Janeiro, Brazil, 8–13 July 2018. [CrossRef]
107. De Santis, E.; Rizzi, A.; Sadeghian, A. Hierarchical genetic optimization of a fuzzy logic system for energy flows management in microgrids. *Appl. Soft Comput. J.* **2017**. [CrossRef]
108. Venayagamoorthy, G.K.; Sharma, R.K.; Gautam, P.K.; Ahmadi, A. Dynamic Energy Management System for a Smart Microgrid. *IEEE Trans. Neural Netw. Learn. Syst.* **2016**. [CrossRef]
109. Ma, L.; Liu, N.; Zhang, J.; Tushar, W.; Yuen, C. Energy Management for Joint Operation of CHP and PV Prosumers Inside a Grid-Connected Microgrid: A Game Theoretic Approach. *IEEE Trans. Ind. Inform.* **2016**. [CrossRef]
110. Jia, K.; Chen, Y.; Bi, T.; Lin, Y.; Thomas, D.; Sumner, M. Historical-Data-Based Energy Management in a Microgrid with a Hybrid Energy Storage System. *IEEE Trans. Ind. Inform.* **2017**. [CrossRef]
111. Arcos-Aviles, D.; Pascual, J.; Guinjoan, F.; Marroyo, L.; Sanchis, P.; Marietta, M.P. Low complexity energy management strategy for grid profile smoothing of a residential grid-connected microgrid using generation and demand forecasting. *Appl. Energy* **2017**. [CrossRef]
112. Aldaouab, I.; Daniels, M.; Hallinan, K. Microgrid cost optimization for a mixed-use building. In Proceedings of the 2017 IEEE Texas Power and Energy Conference (TPEC), College Station, TX, USA, 9–10 February 2017. [CrossRef]
113. Liu, N.; Yu, X.; Wang, C.; Wang, J. Energy Sharing Management for Microgrids with PV Prosumers: A Stackelberg Game Approach. *IEEE Trans. Ind. Inform.* **2017**. [CrossRef]
114. Nwulu, N.I.; Xia, X. Optimal dispatch for a microgrid incorporating renewables and demand response. *Renew. Energy* **2017**. [CrossRef]
115. Astaneh, M.; Roshandel, R.; Dufo-López, R.; Bernal-Agustín, J.L. A novel framework for optimization of size and control strategy of lithium-ion battery based off-grid renewable energy systems. *Energy Convers. Manag.* **2018**. [CrossRef]
116. Neves, D.; Pina, A.; Silva, C.A. Comparison of different demand response optimization goals on an isolated microgrid. *Sustain. Energy Technol. Assess.* **2018**. [CrossRef]
117. Wei, Q.; Liu, D.; Lewis, F.L.; Liu, Y.; Zhang, J. Mixed Iterative Adaptive Dynamic Programming for Optimal Battery Energy Control in Smart Residential Microgrids. *IEEE Trans. Ind. Electron.* **2017**. [CrossRef]
118. Yan, B.; Luh, P.B.; Warner, G.; Zhang, P. Operation and Design Optimization of Microgrids with Renewables. *IEEE Trans. Autom. Sci. Eng.* **2017**. [CrossRef]
119. Akter, M.N.; Mahmud, M.A.; Oo, A.M.T. A hierarchical transactive energy management system for microgrids. In Proceedings of the 2016 IEEE Power and Energy Society General Meeting (PESGM), Boston, MA, USA, 17–21 July 2016. [CrossRef]
120. Basaran, K.; Cetin, N.S.; Borekci, S. Energy management for on-grid and off-grid wind/PV and battery hybrid systems. *IET Renew. Power Gener.* **2016**. [CrossRef]

121. Lai, C.S.; Jia, Y.; Xu, Z.; Lai, L.L.; Li, X.; Cao, J.; McCulloch, M.D. Levelized cost of electricity for photovoltaic/biogas power plant hybrid system with electrical energy storage degradation costs. *Energy Convers. Manag.* **2017**. [CrossRef]
122. Muñoz Maldonado, Y.A. Optimización de Recursos Energéticos en Zonas Aisladas Mediante Estrategias de Suministro y Consumo. Ph.D. Thesis, Universitat Politècnica de València, València, Spain, 2012. [CrossRef]

© 2019 by the authors. Licensee MDPI, Basel, Switzerland. This article is an open access article distributed under the terms and conditions of the Creative Commons Attribution (CC BY) license (http://creativecommons.org/licenses/by/4.0/).

Article

Development of a DSP Microcontroller-Based Fuzzy Logic Controller for Heliostat Orientation Control

Eugenio Salgado-Plasencia, Roberto V. Carrillo-Serrano and Manuel Toledano-Ayala *

División de Investigación y Posgrado, Facultad de Ingeniería, Universidad Autónoma de Querétaro, Cerro de las Campanas s/n, Santiago de Querétaro, Querétaro 76010, Mexico; eugenio.salgado@uaq.mx (E.S.-P.); roberto.carrillo@uaq.mx (R.V.C.-S.)
* Correspondence: toledano@uaq.mx; Tel.: +52-442-144-5820

Received: 25 December 2019; Accepted: 17 February 2020; Published: 28 February 2020

Abstract: This paper describes the design and implementation of a heliostat orientation control system based on a low-cost microcontroller. The proposed system uses a fuzzy logic controller (FLC) with the Center of Sums defuzzification method embedded on a dsPIC33EP256MU806 Digital Signal Processor (DSP), in order to modify the orientation of a heliostat by controlling the angular position of two DC motors connected to the axes of the heliostat. The FLC is compared to a traditional Proportional-Integral-Derivative (PID) controller to evaluate the performance of the system. Both the FLC and PID controller were designed for the position control of the heliostat DC motors at no load, and then they were implemented in the orientation control of the heliostat using the same controller parameters. The experimental results show that the FLC has a better performance and flexibility than a traditional PID controller in the orientation control of a heliostat.

Keywords: heliostat; sun tracking; solar energy; embedded system; fuzzy logic control; center of sums defuzzification method

1. Introduction

The output power produced by a solar plant is proportional to the amount of solar energy absorbed by the system. Therefore, a sun tracking system (STS) with a high degree of accuracy is necessary to avoid losses in the output power of solar plants. STSs are usually classified into two categories [1]: passive sun tracking systems, which use the expansion of a gas caused by the solar radiation to move the mechanical structure of the tracker, and active sun tracking systems, which use motors, gears and electric controllers to drive concentration and absorption devices in a solar plant. There are two types of active sun tracking systems based on their controlling methods [2]: sensor driver systems (SDSs) and microprocessor driver systems (MDSs). SDSs use photosensors in order to detect a change in light sources and convert it into an electrical signal, which is used to obtain the position of the sun. However, there are tracking errors when the sensors cannot produce an electrical signal due to low solar radiation levels produced by the presence of passing clouds or contamination in the air. MDSs use microprocessors and computer systems to execute mathematical equations based on solar position algorithms and the current date and time to determine the exact position of the sun. MDSs are cheaper than SDSs; however, there is no feedback to verify the position of the sun, and tracking errors may appear due to the precision of the solar position algorithm. Several algorithms with different levels of complexity and accuracy can be found in the literature [3], where the use of a more precise solar position algorithm increases both the accuracy and the computational effort of the system.

Central tower power plants use two-axis sun reflectors called heliostats, which reflect the solar irradiance into a collector tower. Every heliostat has a local control which drives two motors connected to reduction gears, where the trend is to give greater autonomy to the central control by increasing the intelligence of the local control of each heliostat. Additionally, in solar plants, there are changing dynamics due to non-linearities and uncertainties that traditional PID controllers cannot handle. This is because a PID controller may produce high oscillations when it is tightly tuned, and the dynamics of the process varies due to changes in the operating conditions. Hence, the use of more efficient control strategies results in better responses [1]. An FLC is a good alternative to traditional PID controllers, because it can deal with non-linear systems and can be designed by using the knowledge of a human operator without knowing the mathematical model of the system. Although the FLC does not have a better response in time domain than a PID controller, this later one cannot be applied to systems which have a quick change of parameters because it would require to adjust the value of its control gains [4].

FLCs [5–10] and hybrid PID-FLCs [4,11–16] have been applied to control the position of DC motors, showing a good output response and better performance against traditional PID controllers. Furthermore, FLCs have been implemented to control two-axis sun trackers for photovoltaic systems. Yousef [17] was the first to develop a PC-based FLC algorithm to control a two-axis photo-voltaic (PV) solar panel. Afterwards, the FLC has been implemented in different platforms and devices such as microcontrollers [18–20], DSPs [21], personal computers (PCs) [22–25] and field-programmable gate arrays (FPGAs) [26].

Finally, the FLC has also been applied in orientation control of heliostats. Ardehali and Emam [27] performed a comparison between a classical PI and PID controller, a PI-FLC and a PID-FLC for the orientation control of a laboratory-scale heliostat with two mirrors of 0.9 m × 0.7 m, two 15 W DC motors, and a data acquisition system with 20 ms sampling time. The FLC uses three membership functions in order to adjust the PID controller gains. The results showed that PI-FLC presented reductions in the overshoot and better performance than the other controllers. Zeghoudi and Chermitti [28] and Zeghoudi et al. [29] used the Matlab environment in order to simulate the orientation of a heliostat by using an FLC with two different rule bases, comparing the output response with a PI, a PID, a PI-FLC and a neural controller. The results showed a better output response for the FLC compared with the other controllers. Additionally, the FLC with fewer rules showed a better output response to step changes than the FLC with the bigger rule base. Bedaouche et al. [30] simulated the position control of two DC motors in order to modify the orientation of a heliostat by using a PID controller self-adjusted by an FLC. The FLC adjusts each PID controller gain by using an individual rule base of forty-nine rules and the error and change of error values. The results showed a faster output response and a smaller overshoot than a classic PID controller. Jirasuwankul and Manop [31] applied an FLC to control the orientation of a lab-scale heliostat with two stepper motors by using a micro-step driver. The position of the heliostat is obtained by using image processing of the reflected solar radiation on the target. The results showed a good performance of the FLC; however, there are tracking errors when the system cannot process the image due to passing clouds.

Nevertheless, the works cited above have only been presented in simulations and small-scale models. Considering the aforementioned, the objective of this paper is to describe the design and implementation of a two-axis STS for the orientation control of a real-scale azimuth-elevation heliostat by using an FLC implemented on a low-cost microcontroller-based embedded system. The comparison between the FLC and a PID controller has also been done.

2. Heliostat Orientation Control

The orientation control system is presented in Figure 1. The control system modifies the angular position of two DC motors connected to the axes of the heliostat through two worm drive mechanisms to guide the heliostat to the desired position. A microcontroller unit (MCU) calculates the position of the sun and the desired angles of the heliostat in order to reflect the solar radiation on a specific target by using the geographic position of the heliostat and the current time and date values. Afterwards, a position control algorithm calculates the error between the desired and current angular position of the heliostat axes by using two rotary encoders in order to obtain a control signal which orients the heliostat by using two motor drivers that allow the bidirectional control of the DC motors.

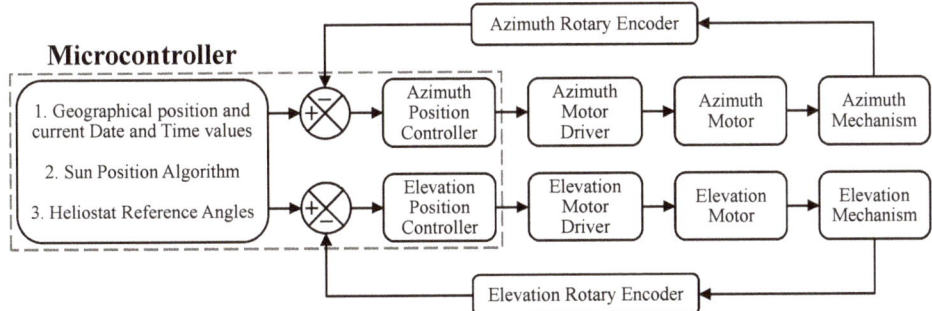

Figure 1. Orientation control diagram.

2.1. DC Motor Mathematical Model

A DC motor can be described by using the equivalent model shown in Figure 2. The reduced transfer function of the armature-controlled DC motor is given by (1) [32].

$$G(s) = \frac{\Theta(s)}{V_a(s)} = \frac{k_m}{s[R_a(Js+b)+k_bk_m]} \quad (1)$$

where k_m represents the motor torque constant and k_b represents the back electromotive-force constant.

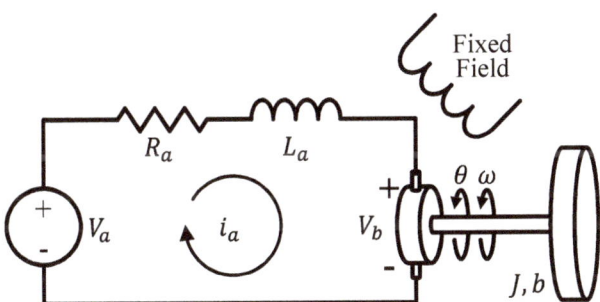

Figure 2. Equivalent model of a permanent magnet brushed DC motor.

The mathematical model of the DC motor can be estimated by using the step signal response method with the motor speed response under a fixed voltage. The transfer function of the position and speed model can be described by (2) and (3).

$$\frac{\Theta(s)}{V_a(s)} = \frac{\frac{k_m}{JR_a}}{s\left(s+\frac{bR_a+k_bk_m}{JR_a}\right)} = \frac{C_k}{s(s+C_p)} \quad (2)$$

$$\frac{\Omega(s)}{V_a(s)} = \frac{C_k}{s+C_p} = \frac{\frac{C_k}{C_p}}{\frac{1}{C_p}s+1} = \frac{K}{\tau s+1} \quad (3)$$

where C_k and C_p are fixed parameters, τ represents the time constant, and K represents the steady-state gain of the system.

The steady-state gain is the ratio of the output and the input in steady-state [33] and is given by (4).

$$K = \frac{\omega_s}{u_{step}} = \frac{C_k}{C_p} \quad (4)$$

where ω_s represents the steady speed of the DC motor and u_{step} represents the step input signal.

Finally, the transfer function of the DC motor is given by (5).

$$G(s) = \frac{\Theta(s)}{V_a(s)} = \frac{\frac{\omega_s}{u_{step}}}{s\left(s + \frac{1}{\tau}\right)} \qquad (5)$$

2.2. Control Algorithms

2.2.1. PID Controller

The PID controller is the most commonly used in industrial applications due to its simple structure. However, its linear nature makes it not very suitable for non-linear systems. It is a control technique which reduces the error ($e(t)$) of a system using three control gains (Proportional, Integral and Derivative) in a mathematical operation to produce a control output ($u(t)$). The equation for the PID controller in the time domain is described by (6) [32].

$$u(t) = K_p e(t) + K_i \int e(t)dt + K_d \frac{de(t)}{dt} \qquad (6)$$

When the controller is digital, it can be approximated with a backward difference and a sum for the derivative and integral terms [4], respectively. The digital PID controller is given by (7).

$$u(n) = K_p e(n) + K_i \sum_{j=1}^{n} e(j) T_s + K_d \frac{e(n) - e(n-1)}{T_s} \qquad (7)$$

where n and T_s represent the number of the sample and the sample time of the digital system.

2.2.2. Fuzzy Logic Controller

An FLC uses the experience of an expert instead of the mathematical model of the system to control a plant, and it can deal with complex non-linear systems with unknown mathematical models. The controller produces a control signal using four blocks [34]: fuzzification, inference engine, rule base and defuzzification, as shown in Figure 3.

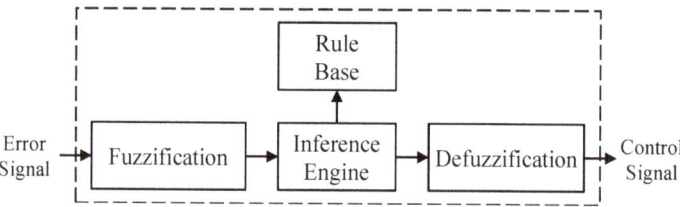

Figure 3. Components of the FLC.

The FLC is graphically shown in Figure 4. The fuzzification module converts the input values into fuzzy sets using the singleton fuzzification, which evaluates the membership value of the input value. The inference mechanism determines the values of the output fuzzy sets by using an "if–then" rule base, which describes the relationship between the input and output variables based on their linguistic terms. In Mamdani fuzzy systems, the rule base determines the output fuzzy set value taking the minimum value of the combination of two or more input fuzzy set values as a consequence of one rule in the rule base. Finally, the defuzzification module gets a scalar value by combining the scaled output fuzzy sets values.

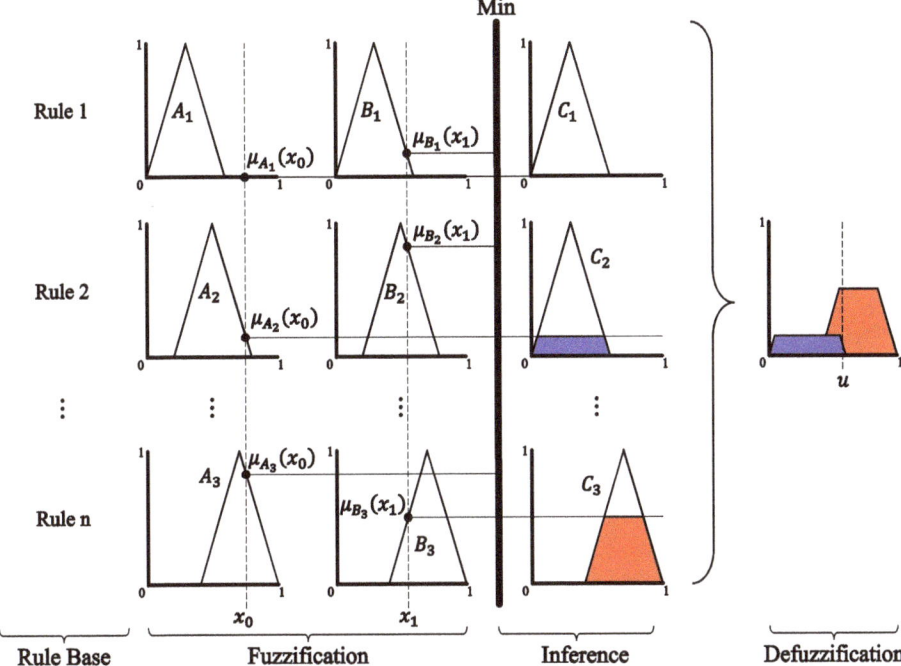

Figure 4. Fuzzy logic controller algorithm.

The reduction of the number of fuzzy rules as long as they express a similar relationship decreases the computational effort and memory requirements used in the implementation of the controller [35]. Therefore, it is necessary to eliminate the less critical rules in order to obtain faster controller actions [36].

The defuzzification module is another component that can be modified in order to obtain a fast response of the controller. There are many defuzzification methods proposed in the literature [34]. The center of gravity method (CoG), also called the center of area method (CoA), is the most widely used of all the defuzzification methods. Nonetheless, this method has a very high computational effort. The CoG method calculates the area under the combined output fuzzy sets by sampling them between the minimum and maximum values of the output fuzzy sets, as shown in Figure 5a. The drawback of the CoG is that it requires more samples to obtain a more accurate output value. The output value of the CoG defuzzification method is given by (8).

$$u = \frac{\sum\limits_{i=1}^{n} \mu(x_i) x_i}{\sum\limits_{i=1}^{n} x_i} \quad , \tag{8}$$

where x_i represents a value between the minimum and maximum values of the scaled output fuzzy sets, and n represents the number of the samples.

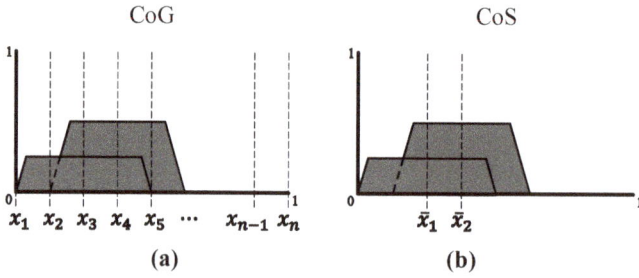

Figure 5. Center of Gravity (**a**) and Center of Sums (**b**) defuzzification methods.

Another defuzzification method is the center of sums method (CoS), which is a fast method because of its computational simplicity [37]. This method calculates the average between the centroid and the area of each scaled output fuzzy set. The drawback of the CoS is that the intersecting areas are added twice, as shown in Figure 5b. The output value of the CoS defuzzification method is given by (9).

$$u = \frac{\sum_{i=1}^{n} \mu(\bar{x}_i) A_i}{\sum_{i=1}^{n} A_i} \qquad (9)$$

where $\bar{x}_i$ and A_i represent the centroid and the scaled area of the output fuzzy set i, and n represents the number of the output fuzzy sets.

2.3. Sun Position and Heliostat Angles

Due to the fact that the relative position of the sun in the sky changes throughout the day, it is necessary to use a solar tracker in order to know the location of the sun at any time. The position of the sun with respect to the observer can be described by a reference system of horizontal coordinates using two angles: the azimuth angle and the elevation angle [1]. The angles of the solar vector $\vec{S}$ are denoted by A_s and E_s, respectively, as shown in Figure 6a. The azimuth angle is measured in relation to the South (0°), and it is negative to the East (−90°) and positive to the West (90°). The elevation angle of the sun ranges from the horizon (0°) to the zenith (90°).

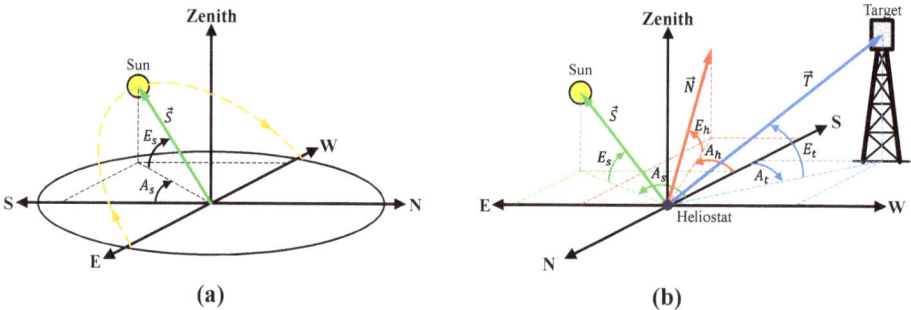

Figure 6. Solar vector (**a**) and vectors and angles of the heliostat (**b**).

For the heliostat to reflect the solar irradiance towards the central receiver, the heliostat surface normal vector $\vec{N}$ must be the bisector of the angle formed by the fixed vector pointing to the receiver from the reflective surface of the heliostat $\vec{T}$ and the solar vector [1] (Figure 6b). The azimuth and elevation angles of the solar vector are given by the Grena [38] algorithm, which has a maximum error of 0.0027°. The solar position algorithm takes the geographical coordinates of the heliostat and the current date and time of the day as input data. The algorithm also uses the monthly average local values of temperature and atmospheric pressure to calculate the atmospheric refraction correction of the elevation angle of the sun. The azimuth and elevation angles of the target vector are obtained by using spherical coordinates. The normal vector is obtained by the addition of the unit vectors of the solar and target vectors.

$$\vec{N} = \left(\hat{S}_x + \hat{T}_x, \quad \hat{S}_y + \hat{T}_y, \quad \hat{S}_z + \hat{T}_z \right) \tag{10}$$

where $\hat{S}$ and $\hat{T}$ are given by (11) and (12).

$$\hat{S} = \left(\sin(A_s)\cos(E_s), \quad \cos(A_s)\cos(E_s), \quad \sin(E_s) \right) \tag{11}$$

$$\hat{T} = \left(\sin(A_t)\cos(E_t), \quad \cos(A_t)\cos(E_t), \quad \sin(E_t) \right) \tag{12}$$

Finally, the azimuth and the elevation angles of the normal vector are given by (13) and (14).

$$A_h = \tan^{-1}\left(\frac{\vec{N}_y}{\vec{N}_x} \right) \tag{13}$$

$$E_h = \tan^{-1}\left(\frac{\vec{N}_z}{\sqrt{\vec{N}_x^2 + \vec{N}_y^2}} \right) \tag{14}$$

2.4. Embedded System

The block diagram of the embedded system is shown in Figure 7. The heliostat orientation control is implemented in a dsPIC33EP256MU806 MCU running with a clock frequency of 48 MHz. The current date and time values are given by a real-time clock (RTC) model DS1307 with I2C (Inter-Integrated Circuit) serial interface protocol. Two H-Bridge motor drivers built with four bipolar junction transistors (tip135 and tip136) are connected to the embedded system in order to change the direction of rotation of the DC motors by using an external power supply and two control signals from the MCU for each DC motor, whereas the feedback signal of the controller is given by two single-turn absolute rotary encoders model CAS60RS12A10SGG with synchronous serial interface (SSI) protocol and 12 bits of resolution (4096 pulses per revolution). Both rotary encoders are connected to the axes of the heliostat in order to obtain the real position of the heliostat. The system also contains an alphanumeric LCD Display to visualize the initial controller parameters, an analog thumb joystick for the manual heliostat control, a UART block to send data to a computer to perform graphical analysis, and a programming port ICSP (In-Circuit Serial Programming).

The algorithm of the embedded system was designed and developed by using CCS C Compiler software and is shown in Figure 8. All fixed values are read from a database at the start of the program. These values include the geographical position of the heliostat, distance to the target, local weather record, configuration data of the microcontroller peripherals, and parameters and grogram functions of the control algorithms. Afterwards, the program runs an infinite loop and waits for a start command to move the heliostat to the desired position. The program uses four 16-bit timers to generate software interrupts at fixed intervals of time in order to operate different components of the system. Timer1 generates a 200 kHz frequency square signal in order to communicate the MCU with the rotary encoders through the SSI protocol, timer2 establishes the period of the PWM signal which controls the speed and position of the DC motors by using the motor drivers, timer3 performs a software interrupt every 10 ms for the sampling time of the control algorithms, and timer4 performs a software interrupt every second in order to send the data to the UART block and read the time and date values from the RTC.

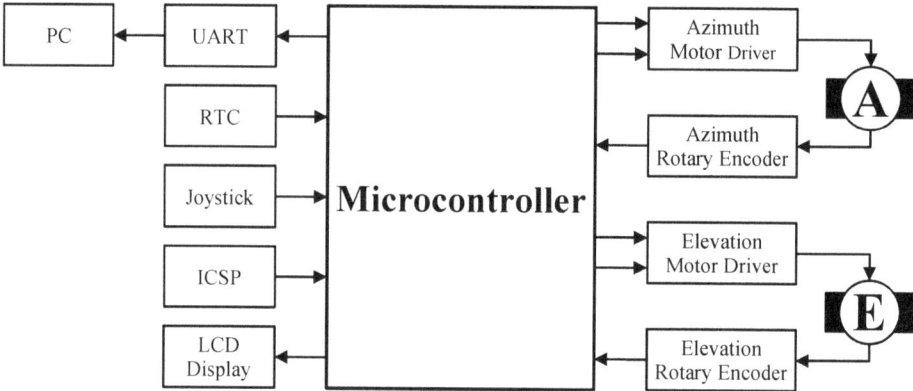

Figure 7. Block diagram of the embedded system.

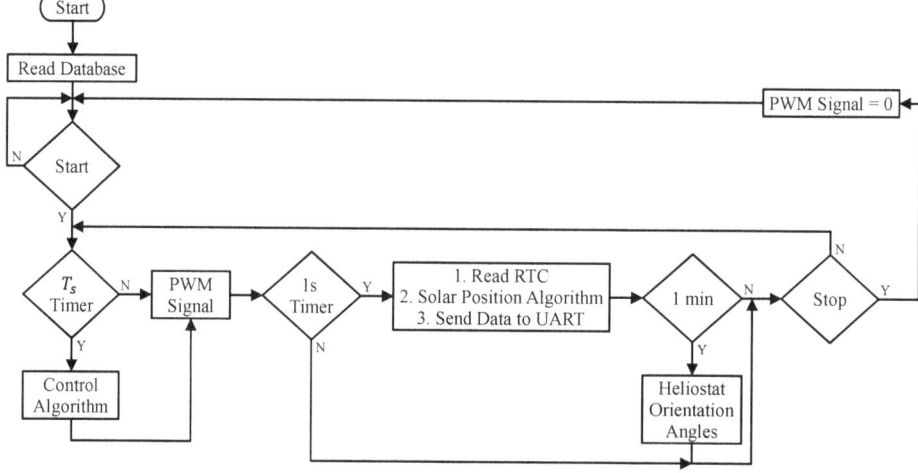

Figure 8. Embedded system algorithm.

Once the start command is received, the program reads the date and time values from the RTC to compute the position of the sun by using the solar position algorithm. Afterwards, the program calculates the desired position of the heliostat to determine the reference values of the control algorithm. Finally, the program calculates the value of the error between the reference values and the position of the heliostat axes which is given by the rotary encoders and determines the control signal of each DC motor by using a program function that takes the error value and returns the values of the voltage that must be supplied to each DC motor. The voltage values are converted into duty cycle values of the PWM signals, which are supplied to the motor drivers in order to move the heliostat to the desired position by adjusting the angular position of each DC motor.

The position of the sun is calculated every second when the value of the RTU changes. However, the reference values of the control algorithm can be set in a fixed period without producing a significant error in the incidence of solar irradiance in the target. Therefore, the desired position of the heliostat is calculated every minute.

The control algorithms are shown in Figures 9–11, for the PID controller and FLC with the CoG and Cos defuzzification methods, respectively.

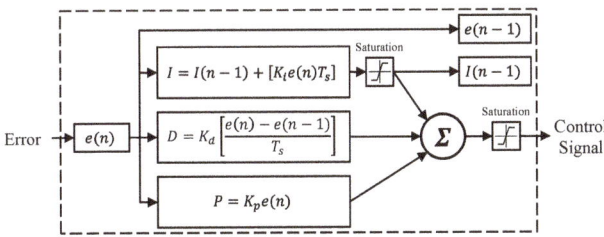

Figure 9. PID controller algorithm.

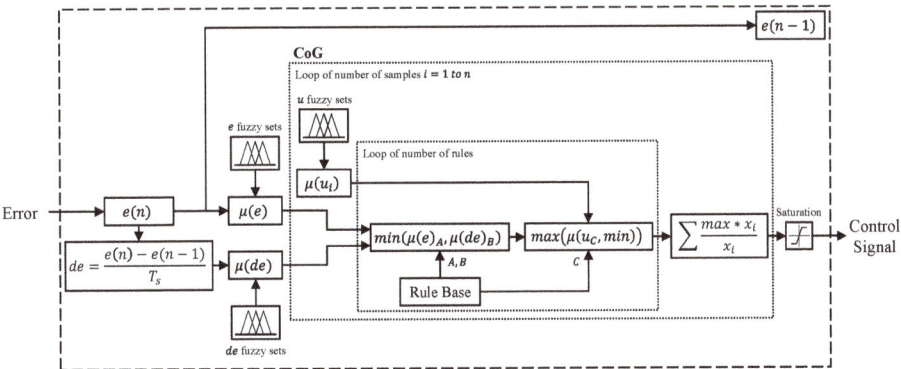

Figure 10. Fuzzy logic controller algorithm with the center of gravity defuzzification method.

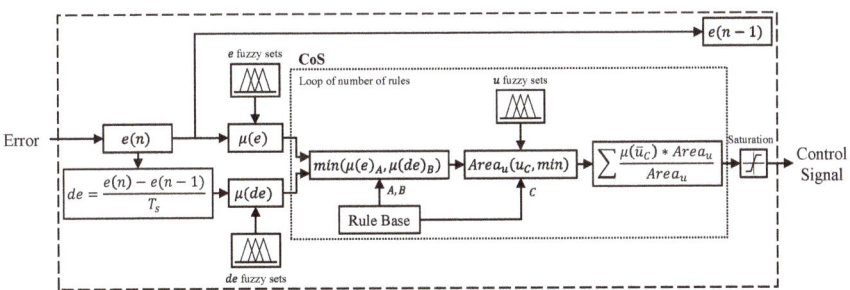

Figure 11. Fuzzy logic controller algorithm with the center of sums defuzzification method.

The PID controller uses the error value to obtain the control signal by using the control gains loaded from the database and Equation (7). A saturation block is used on the integral term to limit its value and obtain a faster response at changes in the error value.

The FLC algorithm obtains the value of the change of error by using the error value and a backward difference, in order to evaluate the input fuzzy sets. Afterwards, the rule base determines the output fuzzy set that corresponds to the values of the error and changes of error and combines it according to the defuzzification method in order to obtain the control signal. The CoG defuzzification method executes a loop for the number of samples that evaluates the output fuzzy sets. In each iteration of the loop, all the rules are evaluated by using another loop for the number of rules in order to obtain the maximum value of the evaluated output fuzzy sets in the sample value, as shown in Figure 5a. Finally, all resulting values are added to obtain the output value by using Equation (8). The CoS defuzzification method only executes one loop, calculating the output value by using values of the scaled area and the centroid of each output fuzzy set, as indicated in Equation (9). The values of the centroid of the output fuzzy sets are calculated once at the beginning of the program and do not change.

The values of the error and integral term are saved in order to calculate the terms used in the next sample of the control algorithms. There is also a saturation block to limit the output signal of the control algorithms at the rated voltage value of the DC motors.

2.4.1. Controller Parameters

The algorithms of the FLC and PID controller were designed for the position control of the DC motors at no load. Both controllers were tuned to accomplish with the design parameters of 10 ms sampling time and 100 ms of rising time without overshoot for the smallest change in the reference signal in order to reduce the energy consumed by the DC motors when the heliostat is moving [27].

Figure 12 shows the block diagram of the FLC. It is a two-input and one-output controller, three fuzzy sets in each input and output signal, a rule base with nine "if–then" rules, a Mamdani inference engine and the CoS defuzzification method. Additionally, there are two processing blocks due to the difference between the fuzzy sets values and the values of the input and output signals. The processing values are given by (15)–(17).

$$e* = \frac{e}{\pi} \tag{15}$$

$$de* = \frac{de T_s}{\pi} \tag{16}$$

$$u* = u V_{max} \tag{17}$$

where $e*$, $de*$ and $u*$ represent the processing values of the input and output signals, and V_{max} represents the maximum voltage signal of the DC motors.

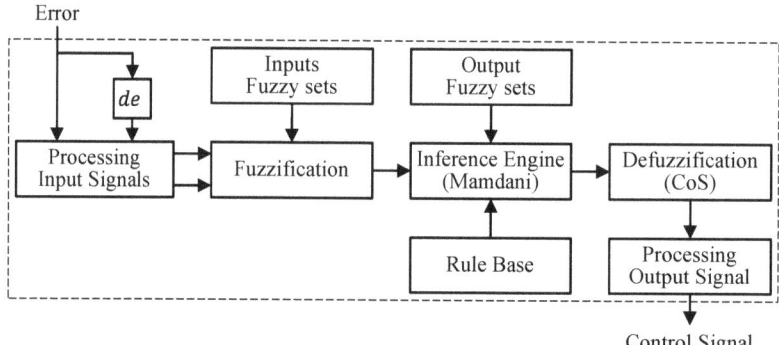

Figure 12. Block diagram of the fuzzy logic controller.

The values of the fuzzy sets and the rule base of the FLC are shown in Figure 13 and Table 1, where the negative, middle, and positive values are denoted by the linguistic variables N, Z and P, respectively. The number and values of the fuzzy sets were selected in order to the control signal of the FLC can modify the position of the DC motor due to the smallest change of the error with a low computational effort. The symmetric shape of the fuzzy sets allows the controller to modify the direction of rotation of the DC motors with the same amplitude of the control signal, which corresponds to the values of the error and change of error.

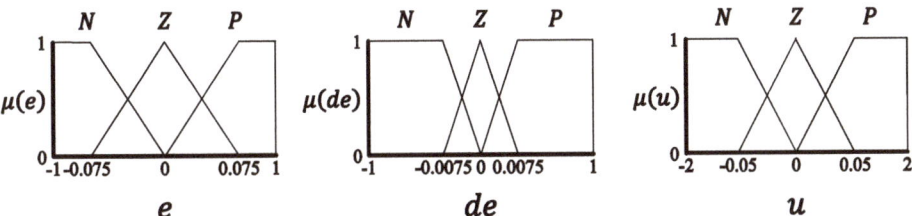

Figure 13. Fuzzy sets of the fuzzy logic controller.

Table 1. Rule base of the fuzzy logic controller.

de \ e	N	Z	P
N	N	N	Z
Z	N	Z	P
P	Z	P	P

The resulted control surface of the FLC is presented in Figure 14, showing the relationship between the inputs and the output as a consequence of the values of the fuzzy sets, the if–then rule base, and the CoS defuzzification method. The output value of the defuzzification varies from −1 to 1; therefore, using Equation (17), the DC motor supply voltage ranges from $-V_{max}$ to V_{max}.

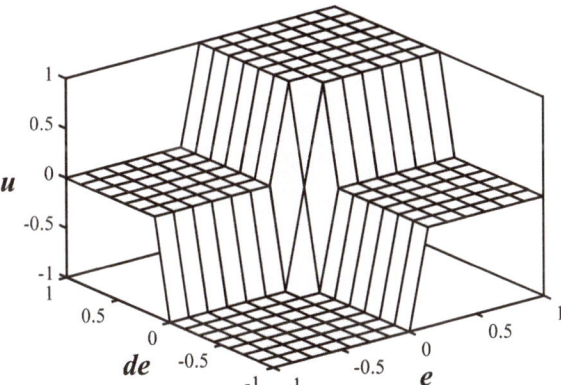

Figure 14. FLC control surface.

For the PID controller, the transfer function of the DC motor is estimated by using the step signal response method in order to obtain the control gains of the PID controller to comply with the design parameters. The angular velocity can be approximated by using a discrete derivative term, as shown in Equation (7). The angular velocity is given by (18).

$$\omega(n) = \frac{\theta(n) - \theta(n-1)}{T_s} \tag{18}$$

The step response of the DC motor and the transfer function parameters are shown in Figure 15, where $\omega = 0.5369 \frac{rad}{s} = 5.126$ rpm is approximately the rated speed reported in the DC motor datasheet, as shown in Table 3.

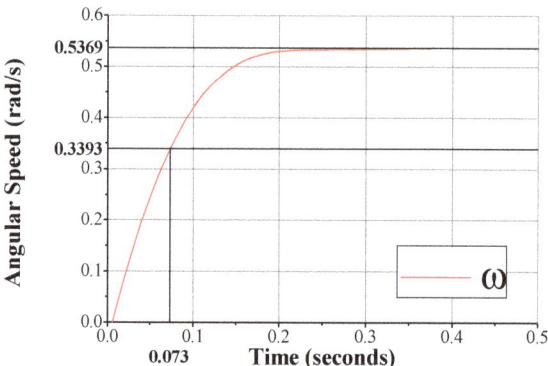

Figure 15. Step response of the DC motor.

The obtained mathematical model of the DC motor at no load is described by (19).

$$G_M(s) = \frac{\Theta_M(s)}{V_{aM}(s)} = \frac{0.31275}{s^2 + 13.68925s} \quad (19)$$

Finally, the control gains of the PID controller were obtained by using the Matlab Sisotool Toolbox. A rising time of 0.075 s and an overshoot of less than 5% were chosen as conditions of the output signal of the PID controller in order to accomplish with the design requirements and produce a smooth control signal to reduce the energy consumption for the DC motors. The control gains of the PID controller are shown in Table 2.

Table 2. Control gains of the PID controller for the DC motor at no load.

K_p	K_i	K_d
2250.0	0.025	110.0

The transfer functions of the DC motors connected to the heliostat axes were also obtained with the same method. The mathematical models of the azimuth and elevation axes are described by (20) and (21), respectively.

$$G_A(s) = \frac{\Theta_A(s)}{V_{aA}(s)} = \frac{0.01316}{s^2 + 4.03225s} \quad (20)$$

$$G_E(s) = \frac{\Theta_E(s)}{V_{aE}(s)} = \frac{0.02417}{s^2 + 7.40740s} \quad (21)$$

2.4.2. Setpoint Values

Because of the position of the sun in the sky changes by 1 degree every 4 min, it is not necessary to modify the orientation of the heliostat every second of the day. Therefore, the values of the reference angles are discretized every minute, as shown in Figure 16 for the parameters of Table 4.

To reduce the error due to the resolution of the rotary encoders, the discrete reference value is converted from radians to encoder steps and is rounded to the closest integer value to obtain a final reference value that corresponds to a value in the encoder steps. Therefore, when the heliostat angles reach the desired position, the error signal will be zero. The values of the error between the final and desired reference of the heliostat axes are shown in Figure 17.

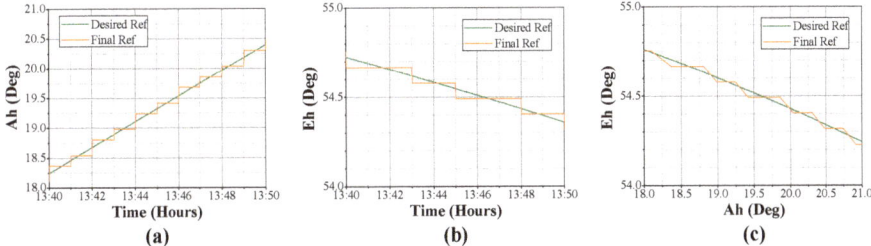

Figure 16. Setpoint values of the orientation control for the azimuth axis (**a**), the elevation axis (**b**) and both axes (**c**).

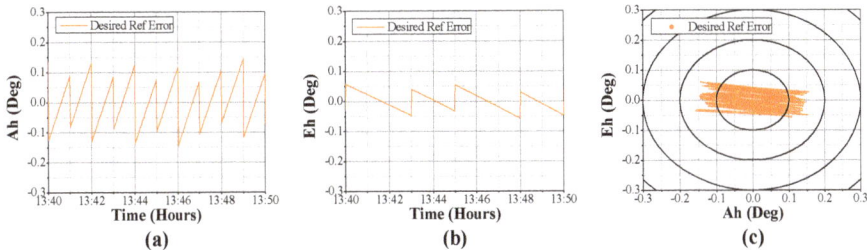

Figure 17. Final reference error values of the orientation control for the azimuth axis (**a**), the elevation axis (**b**) and both axes (**c**).

3. Results and Discussion

The orientation control system was implemented in the heliostat shown in Figure 18, whereas the printed circuit board (PCB) of the embedded system and the motor driver are shown in Figure 19. It is an azimuth–elevation mechanism heliostat, with a worm drive mechanism driven by a DC gear motor model ZYT6590-01 at each axis. The heliostat has a gap which allows directing the facets to the ground. The parameters of the heliostat and the DC motors are presented in Table 3.

Figure 18. Heliostat.

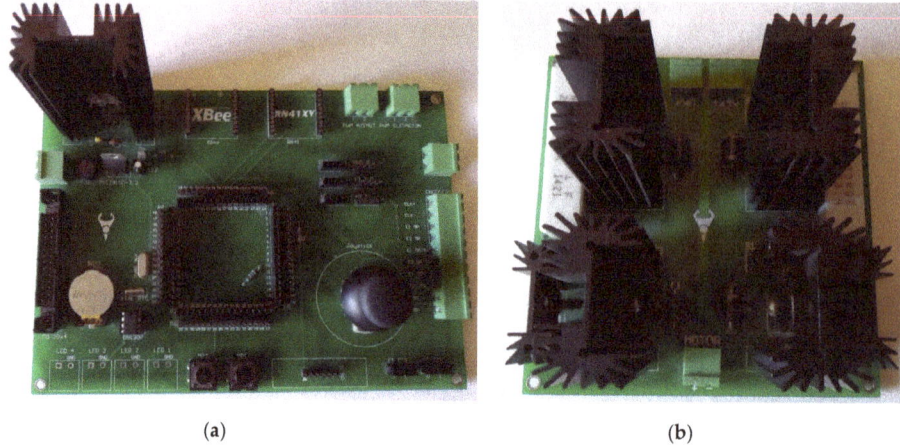

Figure 19. Printed circuit board of the embedded system (**a**) and the motor driver (**b**).

Table 3. Parameters of the heliostat and DC motors.

Parameter	Value	Unit
Total height	5.24	m
Pedestal height	2.85	m
Elevation axis length	4.43	m
Gap between support frames	0.70	m
Number of facets	16	-
Mirror face size	1.2 × 1.2	m
Heliostat mirror area	23	m^2
DC Motors Rated Voltage	24	V
DC Motors Rated Current	≤5	A
DC Motors Rated Torque	100	N·m
DC Motors No Load Speed	5	rpm
DC Motors Gear Ratio	710.5	-

As mentioned already, the control algorithms were designed for the position control of a DC motor at no load. Afterwards, the control algorithms were implemented in the orientation control of the heliostat using the same controller parameters of the position control of the DC motor at no load.

Figure 20 shows the comparison of the consumption time of the PID controller (Figure 20a) and the FLC using the CoS (Figure 20b) and the CoG (Figure 20c) defuzzification methods, where the period of the signals represents the sampling time of the control algorithms. The results show that the FLC with the CoG defuzzification method does not accomplish with the design parameters because of its computational complexity.

The output response of the control algorithms for the position control of a DC motor at no load is shown in Figure 21 for the minimum change in the reference value of 0.087 degrees (1.533 mrad) and a reference value of 180 degrees (π rad). Both control algorithms accomplish with the design parameters for the position control of the DC motor at no load. However, Figure 21d shows that the FLC control signal decreases when the position of the DC motor is reaching the reference value.

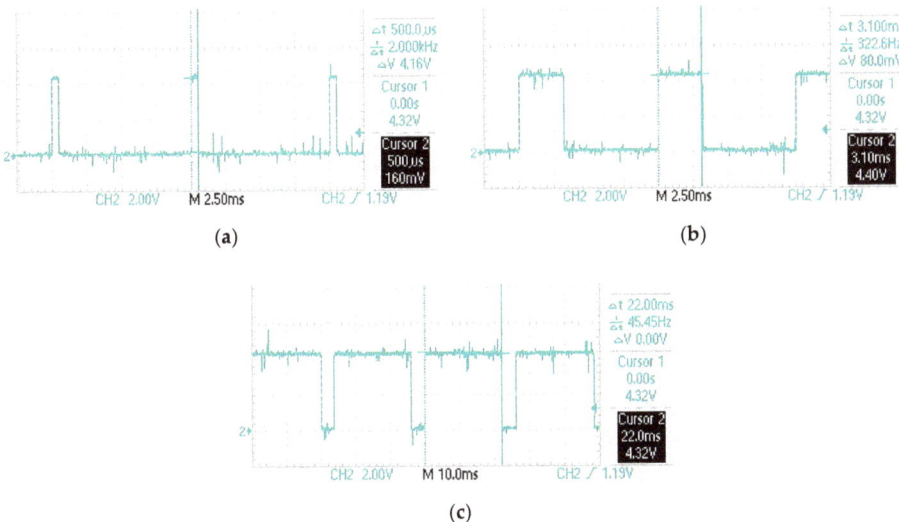

Figure 20. Consumption time of the PID controller (**a**), and the fuzzy logic controller with the CoS (**b**) and the CoG (**c**) defuzzification methods.

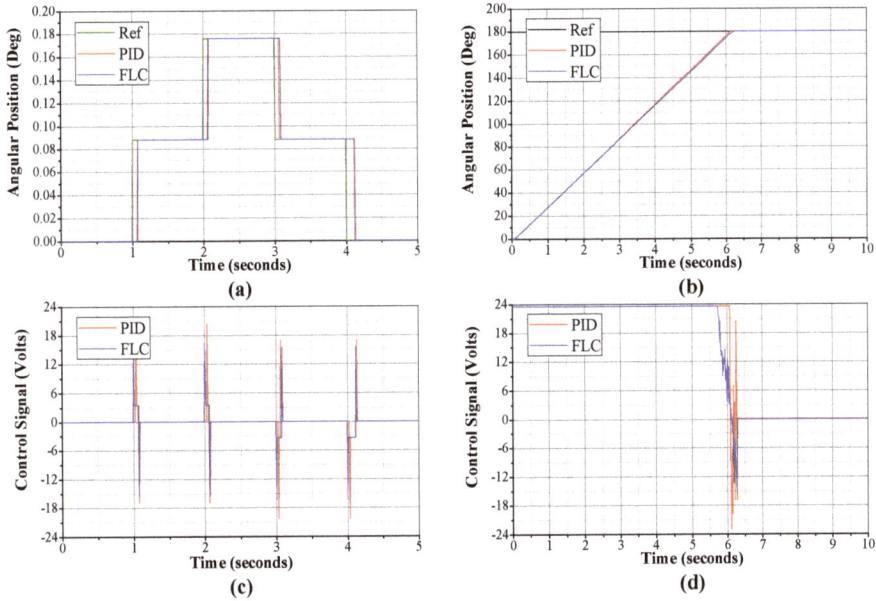

Figure 21. Output response (**a**) and control signal (**c**) of the DC motor at no load at a minimum reference value. Output response (**b**) and control signal (**d**) of the DC motor at no load at a reference value of 180 degrees (π rad).

Finally, Figures 22 and 23 show the output response of the control algorithms for the orientation control for the DC motors at no load and the axes of the heliostat, respectively. The desired angles of the heliostat were calculated using the parameters of Table 4, whereas the error values are shown in Table 5.

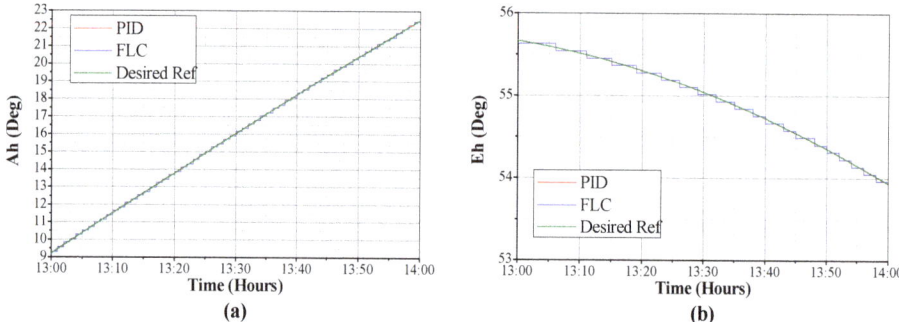

Figure 22. Output response of the orientation control of the DC motors at no load for the azimuth axis (**a**) and the elevation axis (**b**).

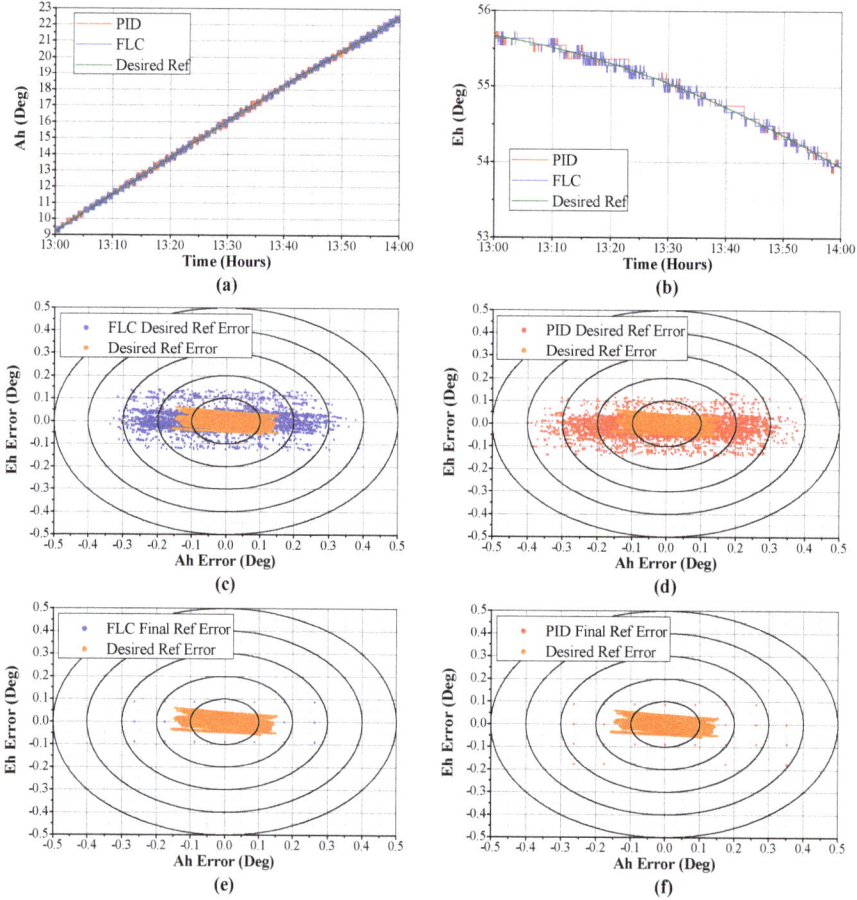

Figure 23. Output response of the orientation control for the azimuth axis (**a**) and the elevation axis (**b**) of the heliostat. Desired reference error values of the orientation control of the heliostat for the fuzzy logic controller (**c**) and the PID controller (**d**). Final reference error values of the orientation control of the heliostat for the fuzzy logic controller (**e**) and the PID controller (**f**).

Table 4. Parameters of the orientation control test.

Parameter	Value
Date	Friday, 13 September 2019
Time	13:00:00–14:00:00
Latitude	20.590636° N
Longitude	100.413226° W
Monthly Mean Atmospheric Pressure	819.795 mbar
Monthly Mean Temperature	20.3 °C
Maximum Wind Speed	8 m/s (28.8 km/h)
Target Height	30.0 m
Heliostat Height	2.85 m
East-West distance to the target	15 m East
North-South distance to the target	35 m North

Table 5. Reference error values of the orientation control.

	Parameter	Final Ref MSE	Desired Ref MSE
DC Motor at no load	PID Azimuth	0.0°	0.068610°
	PID Elevation	0.0°	0.026349°
	FLC Azimuth	0.0°	0.068610°
	FLC Elevation	0.0°	0.026349°
Heliostat	PID Azimuth	0.153941°	0.168669°
	PID Elevation	0.051032°	0.048347°
	FLC Azimuth	0.131647°	0.146435°
	FLC Elevation	0.039328°	0.047251°

The experimental results show a similar Mean Squared Error (MSE) for the orientation control of the DC motors at no load and a similar output response between the orientation control of the heliostat and the final reference value for the FLC (Figure 23a) and the PID controller (Figure 23b), despite the load of the wind over the mechanical structure and the backlash in the axis mechanisms. However, for the orientation control of the heliostat, the FLC shows less dispersed error values (Figure 23c) and smaller final reference error values (Figure 23e) than the PID controller (Figure 23d,f).

4. Conclusions

The orientation control of a heliostat using an FLC was implemented on an embedded system based on a low-cost microcontroller. Also, the comparison against a traditional PID controller was performed. The advantage of the FLC is the fact that it is not necessary to know the mathematical model of the system, because it only uses the experience of an operator, which is easy to incorporate into the controller.

The results show that both controllers exhibit a similar output response for the position control of a DC motor. However, the FLC has a better performance than the PID controller for the orientation control of the heliostat by using the same control parameters for the position control of the DC motor at no load. The FLC has higher flexibility since it is robust in front of changes in the dynamics of the process, whereas for a better output response of the PID controller, the control gains must be tuned for the mathematical models of the heliostat axes. The results also exhibit a smaller MSE of the FLC compared to the PID controller for the orientation control of the heliostat by using only a nine-rule rule base and a fuzzy set of three membership functions in each input and output signal in order to reduce the computational effort of the controller. Additionally, the center of the sums defuzzification method complies with the design parameter of 10 ms sample time, showing a faster response than the center of the gravity defuzzification method.

In a central tower power plant that uses traditional PID controllers for the orientation control of the heliostats, the control gains of the controller of all the heliostats must be adjusted in order to avoid oscillations due to wrong controller parameters tuning. Therefore, the proposed control system can be applied in order to control an entire heliostat field by using the same controller parameters for all the heliostat. The system can also be adjusted to control other sun tracking systems, such as a photovoltaic, solar dish, or parabolic trough systems, which only need the solar tracker system.

Author Contributions: Conceptualization, E.S.-P. and M.T.-A.; Methodology, E.S.-P. and R.V.C.-S.; Project administration, M.T.-A.; Software, E.S.-P.; Validation, E.S.-P. and R.V.C.-S. All authors have read and agreed to the published version of the manuscript.

Funding: This research received no external funding.

Acknowledgments: The authors would like to thank Consejo Nacional de Ciencia y Tecnología (CONACYT-México) for supporting this research.

Conflicts of Interest: The authors declare no conflict of interest.

Abbreviations

CoG	Center of Gravity
CoS	Center of Sums
DSP	Digital Signal Processor
FLC	Fuzzy Logic Controller
FPGA	Field-Programmable Gate Array
LCD	Liquid Crystal Display
MCU	Microcontroller Unit
MDS	Microprocessor Driver System
MSE	Mean Squared Error
PC	Personal Computer
PCB	Printed Circuit Board
PID	Proportional–Integral–Derivative
PV	Photo-Voltaic
PWM	Pulse Width Modulation
RTC	Real-Time Clock
SDS	Sensor Driver System
STS	Sun Tracking System
UART	Universal Asynchronous Receiver-Transmitter
θ	Angular position of the DC motor
ω	Angular velocity of the DC motor
τ	Time constant of the system
K	Steady-state gain of the system
e	Controller error signal
de	Controller change of error signal
u	Controller output signal
T_s	Controller sampling time
V_{max}	Controller maximum output voltage
$\vec{S}$	Solar vector
$\vec{T}$	Target vector
$\vec{N}$	Normal vector of the heliostat
$\hat{S}$	Solar unit vector
$\hat{T}$	Target unit vector
A_s	Solar vector azimuth angle
E_s	Solar vector elevation angle
A_t	Target vector azimuth angle
E_t	Target vector elevation angle
A_h	Heliostat azimuth angle
E_h	Heliostat elevation angle

References

1. Camacho, E.F.; Berenguel, M.; Rubio, F.R.; Martínez, D. *Control of Solar Energy Systems*; Springer: London, UK, 2012; p. 414.
2. Al-Rousan, N.; Isa, N.A.M.; Desa, M.K.M. Advances in solar photovoltaic tracking systems: A review. *Renew. Sustain. Energy Rev.* **2018**, *82*, 2548–2569. [CrossRef]
3. Grena, R. Five new algorithms for the computation of sun position from 2010 to 2110. *Sol. Energy* **2012**, *86*, 1323–1337. [CrossRef]
4. Soufi, N.J. Design and implementation of fuzzy position control system for tracking applications and performance comparison with conventional pid. *IAES Int. J. Artif. Intell. (IJ-AI)* **2012**, *1*, 31–44.
5. Paul-l-Hai, L.; Sentai, H.; Chou, J. Comparison on fuzzy logic and pid controls for a dc motor position controller. In Proceedings of the 1994 IEEE Industry Applications Society Annual Meeting, Denver, CO, USA, 2–6 October 1994; pp. 1930–1935.
6. Bal, G.; Bekiroğlu, E.; Demirbaş, Ş.; Çolak, İ. Fuzzy logic based dsp controlled servo position control for ultrasonic motor. *Energy Convers. Manag.* **2004**, *45*, 3139–3153. [CrossRef]
7. Wang, H.-P. Design of fast fuzzy controller and its application on position control of dc motor. In Proceedings of the 2011 International Conference on Consumer Electronics, Communications and Networks, XianNing, China, 16–18 April 2011; pp. 4902–4905.
8. Chermitti, A.; Zeghoudi, A. A comparison between a fuzzy and pid controller for universal motor. *Int. J. Comput. Appl.* **2014**, *104*, 32–36.
9. Meena, P.K.; Bhushan, B. Simulation for position control of dc motor using fuzzy logic controller. *Int. J. Electron. Electr. Comput. Syst.* **2017**, *6*, 188–191.
10. Rahman, Z.-A.S.A. Design a fuzzy logic controller for controlling position of dc motor. *Int. J. Comput. Eng. Res. Trends* **2017**, *4*, 285–289.
11. Lim, C.M. Implementation and experimental study of a fuzzy logic controller for dc motors. *Comput. Ind.* **1995**, *26*, 93–96. [CrossRef]
12. Ko, J.S.; Youn, M.J. Simple robust position control of bldd motors using integral-proportional plus fuzzy logic controller. *Mechatronics* **1998**, *8*, 65–82. [CrossRef]
13. Pravadalioglu, S. Single-Chip fuzzy logic controller design and an application on a permanent magnet dc motor. *Eng. Appl. Artif. Intell.* **2005**, *18*, 881–890. [CrossRef]
14. Namazov, M.; Basturk, O. Dc motor position control using fuzzy proportional-derivative controllers with different defuzzification methods. *Turk. J. Fuzzy Syst.* **2010**, *1*, 36–54.
15. Natsheh, E.; Buragga, K.A. Comparison between conventional and fuzzy logic pid controllers for controlling dc motors. *Int. J. Comput. Sci. Issues* **2010**, *7*, 128–134.
16. Manikandan, R.; Arulmozhiyal, R. Position control of dc servo drive using fuzzy logic controller. In Proceedings of the 2014 International Conference on Advances in Electrical Engineering (ICAEE), Vellore, India, 9–11 January 2014; pp. 1–5.
17. Yousef, H.A. Design and implementation of a fuzzy logic computer-controlled sun tracking system. In Proceedings of the IEEE International Symposium on Industrial Electronics, Bled, Slovenia, 12–16 July 1999; pp. 1030–1034.
18. Belkasmi, M.; Bouziane, K.; Akherraz, M.; Sadiki, T.; Faqir, M.; Elouahabi, M. Improved dual-axis tracker using a fuzzy-logic based controller. In Proceedings of the 3rd International Renewable and Sustainable Energy Conference (IRSEC), Marrakech, Morocco, 10–13 December 2015; pp. 1–5.
19. Zakariah, A.; Jamian, J.J.; Yunus, M.A.M. Dual-Axis solar tracking system based on fuzzy logic control and light dependent resistors as feedback path elements. In Proceedings of the 2015 IEEE Student Conference on Research and Development (SCOReD), Kuala Lumpur, Malaysia, 13–14 December 2015; pp. 139–144.
20. Toylan, H. Performance of dual axis solar tracking system using fuzzy logic control a case study in Pinarhisar, Turkey. *Eur. J. Eng. Nat. Sci.* **2017**, *2*, 130–136.
21. Ataei, E.; Afshari, R.; Pourmina, M.A.; Karimian, M.R. Design and construction of a fuzzy logic dual axis solar tracker based on dsp. In Proceedings of the 2nd International Conference on Control, Instrumentation and Automation, Shiraz, Iran, 27–29 December 2011; pp. 185–189.
22. Batayneh, W.; Owais, A.; Nairoukh, M. An intelligent fuzzy based tracking controller for a dual-axis solar pv system. *Autom. Constr.* **2013**, *29*, 100–106. [CrossRef]

23. Baran, N.; Sinha, D. Fuzzy logic-based dual axis solar tracking system. *Int. J. Comput. Appl.* **2016**, *155*, 13–18.
24. Huang, C.-H.; Pan, H.-Y.; Lin, K.-C. Development of intelligent fuzzy controller for a two-axis solar tracking system. *Appl. Sci.* **2016**, *6*, 130. [CrossRef]
25. Zeghoudi, A.; Hamidat, A.; Takilalte, A.; Debbache, M. Contribution to the control of a tracker solar using hybrid controller and artificial intelligence systems. In Proceedings of the 4th International Seminar on New and Renewable Energies, Ghardaïa, Algeria, 24–25 October 2016; pp. 1–9.
26. Benzekri, A.; Azrar, A. FPGA-based design process of a fuzzy logic controller for a dual-axis sun tracking system. *Arab. J. Sci. Eng.* **2014**, *39*, 6109–6123. [CrossRef]
27. Ardehali, M.M.; Emam, S.H. Development, design and experimental testing of fuzzy-based controllers for a laboratory scale sun-tracking heliostat. *Fuzzy Inf. Eng.* **2011**, *3*, 247–257. [CrossRef]
28. Zeghoudi, A.; Chermitti, A. Speed control of a dc motor for the orientation of a heliostat in a solar tower power plant using artificial intelligence systems (flc and nc). *Res. J. Appl. Sci. Eng. Technol.* **2015**, *10*, 570–580. [CrossRef]
29. Zeghoudi, A.; Chermitti, A.; Benyoucef, B. Contribution to the control of the heliostat motor of a solar tower power plant using intelligence controller. *Int. J. Fuzzy Syst.* **2015**, *18*, 741–750. [CrossRef]
30. Bedaouche, F.; Gama, A.; Hassam, A.; Khelifi, R.; Boubezoula, M. Fuzzy pid control of a dc motor to drive a heliostat. In Proceedings of the 2017 International Renewable and Sustainable Energy Conference, Tangier, Morocco, 4–7 December 2017; pp. 1–6.
31. Jirasuwankul, N.; Manop, C. A lab-scale heliostat positioning control using fuzzy logic based stepper motor drive with micro step and multi-frequency mode. In Proceedings of the 2017 IEEE International Conference on Fuzzy Systems (FUZZ-IEEE), Naples, Italy, 9–12 July 2017; pp. 1–6.
32. Dorf, R.C.; Bishop, R.H. *Modern Control Systems*, 12th ed.; Pearson Education, Inc.: Upper Saddle River, NJ, USA, 2011.
33. Aguado, A.B.; Martínez, M.I. *Identificación y Control Adaptativo*; Pearson Educación, S.A.: Madrid, Spain, 2003.
34. Lilly, J.H. *Fuzzy Control and Identification*, 1st ed.; John Wiley & Sons, Inc.: Hoboken, NJ, USA, 2010; p. 231.
35. Eminoğlu, İ.; Altaş, İ.H. The effects of the number of rules on the output of a fuzzy logic controller employed to a pm dc motor. *Comput. Electr. Eng.* **1998**, *24*, 245–261. [CrossRef]
36. Abdalla, M.; Al-Jarrah, T. Optimal fuzzy controller: Rule base optimized generation. *Control. Eng. Appl. Inform.* **2018**, *20*, 76–86.
37. Ross, T.J. *Fuzzy Logic with Engineering Applications*, 3rd ed.; John Wiley & Sons Ltd.: Chichester, UK, 2010; p. 585.
38. Grena, R. An algorithm for the computation of the solar position. *Sol. Energy* **2008**, *82*, 462–470. [CrossRef]

© 2020 by the authors. Licensee MDPI, Basel, Switzerland. This article is an open access article distributed under the terms and conditions of the Creative Commons Attribution (CC BY) license (http://creativecommons.org/licenses/by/4.0/).

Article
Adaptive Pitch Control of Variable-Pitch PMSG Based Wind Turbine

Jian Chen [1], Bo Yang [2,*], Wenyong Duan [1], Hongchun Shu [2], Na An [2], Libing Chen [1] and Tao Yu [3]

1. School of Electrical Engineering, Yancheng Institute of Technology, Yancheng 224051, China; cjycit@163.com (J.C.); dwy1985@126.com (W.D.); cumtclb@126.com (L.C.)
2. Faculty of Electric Power Engineering, Kunming University of Science and Technology, Kunming 650500, China; yangboffg@sina.com (H.S.); Anna073000@163.com (N.A.)
3. College of Electric Power, South China University of Technology, Guangzhou 510640, China; taoyu1@scut.edu.cn
* Correspondence: yangbo_ac@outlook.com; Tel.: +86-18314596103

Received: 29 August 2019; Accepted: 24 September 2019; Published: 1 October 2019

Abstract: This paper presents an adaptive pitch-angle control approach for a permanent magnet-synchronous generator-based wind turbine (PMSG-WT) connecting with a power grid to limit extracted power above the rated wind speed. In the proposed control approach, a designed perturbation observer is employed for estimating and compensating unknown parameter uncertainties, system nonlinearities, and unknown disturbances. The proposed control approach does not require full state measurements or the accurate system model. Simulation tests verify the effectiveness of the proposed control approach. The simulation results demonstrate that compared with the feedback linearizing controller, conventional vector controller with proportional-integral (PI) loops, and PI controller with gain scheduling, the proposed control approach can always maintain the extracted wind power around rated power, and has higher performance and robustness against disturbance and parameter uncertainties.

Keywords: pitch control; permanent magnet-synchronous generator (PMSG); limit extracted power; nonlinear adaptive control (NAC); perturbation observer

1. Introduction

Wind power generation systems (WPGSs) have become competitive and attractive as exhaustless and clean power sources [1–6]. According to the objectives of variable speed variable-pitch wind turbines (WT), three main operating regions can be observed [7], as illustrated in Figure 1. In Region 1, wind speed is lower than cut-in wind speed (V_{ci}), and the WT does not operate; in Region 2, wind speed is between cut-in wind speed (V_{cut-in}) and rated wind speed (V_{rated}), and the maximum wind power is required to be extracted by rotor speed control; in Region 3, wind speed is between rated wind speed (V_{rated}) and cut-out wind speed ($V_{cut-out}$), and its main control objective is maintaining extracted wind power around rated power via blade pitch control and electromagnetic torque control.

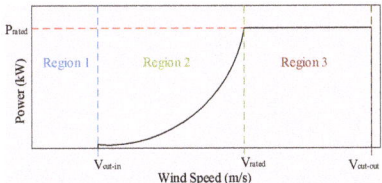

Figure 1. Main operating regions of the wind turbine.

To keep the wind turbine (WT) within its design limits in Region 3, blade pitch and electromagnetic torque control is primarily applied in limiting the extracted wind power [8]. As the electromagnetic torque has much faster response than the mechanical torque, the decoupled control between the WT and generator can be applied [9]. When wind speed is above rated wind speed, for the WT, the control of the mechanical rotation speed is applied to achieve the required pitch angle. The extracted wind power will vary only in proportion to mechanical rotation speed when the mechanical torque keeps at its rated value. Therefore, extracted wind power regulation is entirely dependent upon mechanical rotation speed regulation. A good tracking of a power reference can be achieved while keeping the rotor speed close to its nominal value. The variable of rotor speed reaches large values that can damage the wind turbine behavior performance in rotor speed regulation by pitch controller [10,11]. For the generator, the electromagnetic torque is required to be maintained at its rated value. When the electromagnetic torque or q-axis stator current and mechanical rotation speed are well regulated, the rated mechanical torque can be achieved. Numerous studies have used the linear techniques and designed controllers based on an approximated linear model for pitch-angle control, such as the linear quadratic Gaussian [12], conventional vector control with proportional-integral (PI) loops [13,14] and PI controller with gain scheduling (GSPI) [10,15]. As the WT contains aerodynamic nonlinearities, the linear controllers designed based on a specific operation point cannot obtain satisfactory performance under time-varying wind speed.

To enhance the performance of the conventional VC and LQG, a nonlinear controller is necessary to be designed for the WT pitch control. One effective solution is employing the feedback linearizing control (FLC) approach. The FLC has been widely and successfully applied in solving many practical nonlinear problems [8,16–18]. Compared to the controllers using linear technique and approximated linear model, a better dynamic performance of nonlinear systems can be achieved under the FLC [19]. The FLC provides fully decoupled control of the original nonlinear system and optimal performance for time-varying operation points. In reference work literature [8], an FLC with an Extended Kalman Filter has been successfully applied in the WT control. In the FLC design, full state information is required to be known. Although the FLC provides better performance than the linear quadratic regulator at low wind speeds, no enhanced performance is achieved at high wind speed, because of model uncertainties. The accurate system model is required to be known in the FLC design [20]. To make up these drawbacks of the FLC, robust control [21–23], fuzzy logic control [10,24,25], sliding mode control [26,27], and neural network control [28], have been proposed. Recently, control methods based on observers have been successfully used to reinforce the robustness of disturbances and model uncertainties in power system [29], permanent magnet-synchronous motor [30,31], photovoltaics inverters [32] and WT [33].

In this paper, a nonlinear adaptive controller (NAC) based on observers is investigated for permanent magnet-synchronous generator-based WT (PMSG-WT) to limit the extracted wind power and provide high performance in Region 3. In the designed NAC, it contains one rotor speed controller and two stator current controllers. One third-order states and perturbation observer (SPO), and two second-order perturbation observers (POs) are employed for the estimations of perturbation terms, including parameter uncertainties, coupling nonlinear dynamics, and disturbances of the PMSG-WT. The estimated perturbations are used for compensating the real perturbation and obtaining adaptive linearizing control of the PMSG-WT. The comparisons of simulation studies among the proposed NAC, FLC, VC and GSPI under three different scenarios, e.g., ramp wind speed, random wind speed and field flux variation, are carried out to verify the effectiveness of the proposed NAC.

The remaining parts of this paper is organized as follows. The model of the PMSG-WT is presented in Section 2. Section 3 presents the design of the NAC. In Section 4, simulation studies are carried out for verifying the effectiveness of the proposed NAC in comparing with the FLC, VC and GSPI. Finally, conclusions of this work are presented in Section 5.

2. Model and Problem Formulation

2.1. PMSG-WT Configuration

In Figure 2, a gearless WPGS equipped with a PMSG is connected to the power grid through full-rate back-to-back voltage source converters. Wind power extracted by the WT is transmitted to the direct-driven PMSG. Then, the mechanical power is converted to electrical power by the PMSG. Then, the electrical power is supplied to the power grid via a machine-side converter (MSC) and a grid-side inverter (GSC). The main objective of the MSC is to extract power from wind by controlling the mechanical rotation speed and electromagnetic torque or q-axis stator current, and produce the required stator voltage, whereas the GSC has to enable decoupled control the active and reactive power required by grid codes. The operation control of these two converters can be decoupled by a DC voltage link [16].

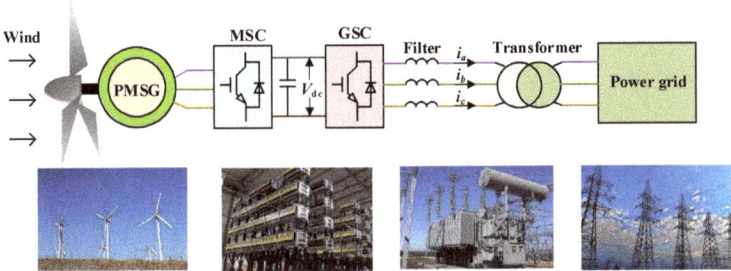

Figure 2. Configuration of a PMSG-WT.

2.2. Aerodynamic Model

The wind power extracted by a WT is represented as [34,35]

$$P_w = \frac{1}{2}\rho \pi R^2 V^3 C_p(\beta, \lambda) \tag{1}$$

$$\lambda = \frac{R\omega_m}{V} \tag{2}$$

where β is the pitch angle, ρ is the air density, V is the wind speed, R is the radius of WT, C_p is the power coefficient, λ is the tip speed ratio, and ω_m is the mechanical rotation speed. The C_p can be defined as a function of β and λ

$$C_p = 0.22\left(\frac{116}{\lambda_i} - 0.4\beta - 5\right)e^{\frac{-12.5}{\lambda_i}} \tag{3}$$

$$\frac{1}{\lambda_i} = \frac{1}{\lambda + 0.08\beta} - \frac{0.035}{\beta^3 + 1} \tag{4}$$

A hydraulic/mechanical actuator can vary the blade pitch. The following first order linear model represents a simplified model of the dynamics:

$$\dot{\beta} = -\frac{\beta}{\tau_\beta} + \frac{\beta_r}{\tau_\beta} \tag{5}$$

where β_r is required pitch angle, and τ_β is the actuator time constant.

The state-space model of the PMSG-WT is given as [35]:

$$\dot{x} = f(x) + g_1(x)u_1 + g_2(x)u_2 + g_3(x)u_3 \tag{6}$$

where

$$f(x) = \begin{bmatrix} -\dfrac{\beta}{\tau_\beta} \\ -\dfrac{R_s}{L_{md}}i_{md} + \dfrac{\omega_e L_{mq}}{L_{md}}i_{mq} \\ -\dfrac{R_s}{L_{mq}}i_{mq} - \dfrac{1}{L_{mq}}\omega_e(L_{md}i_{md} + K_e) \\ \dfrac{1}{J_{tot}}(T_e - T_m - T_f - B\omega_m) \end{bmatrix},$$

$$g_1(x) = [-\dfrac{\beta}{\tau_\beta} \ 0 \ 0 \ 0]^T,$$

$$g_2(x) = [0 \ \dfrac{1}{L_{md}} \ 0 \ 0]^T,$$

$$g_3(x) = [0 \ 0 \ \dfrac{1}{L_{mq}} \ 0]^T,$$

$$x = [\beta \ i_{md} \ i_{mq} \ \omega_m]^T,$$

$$u = [u_1, u_2, u_3]^T = [\beta_r, V_{md}, V_{mq}]^T,$$

$$y = [y_1, y_2, y_3]^T = [h_1(x), h_2(x), h_3(x)]^T = [\omega_m, i_{md}, i_{mq}]^T$$

where $x \in R^4$, $u \in R^3$ and $y \in R^3$ are state vector, input vector and output vector, respectively; $f(x)$, $g(x)$ and $h(x)$ are smooth vector fields. V_{md} and V_{mq} are the d, q axis stator voltages, i_{md} and i_{mq} are the d, q axis stator currents, L_{md} and L_{mq} are d, q axis stator inductances, R_s is the stator resistance, p is the number of pole pairs, K_e is the field flux given by the magnet, J_{tot} is the total inertia of the drive train, B is the friction coefficient of the PMSG, $\omega_e(= p\omega_m)$ is the electrical generator rotation speed, and T_m, T_f and T_e are the WT mechanical torque, static friction torque and electromagnetic torque, respectively.

The electromagnetic torque is expressed as:

$$T_e = p[(L_{md} - L_{mq})i_{md}i_{mq} + i_{mq}K_e] \tag{7}$$

2.3. Pitch Control

To maintain the extracted wind power at rated power in Region 3, it requires that the corresponding pitch angle should be achieved, which in turn requires both the mechanical rotation speed ω_m and the mechanical torque T_m should be kept at their rated values, respectively. The rated mechanical torque T_{mr} is achieved when the electromagnetic torque T_e can track its rated value T_{er} and the ω_m is kept at it rated value. According to Equation (7), the electromagnetic torque T_e can be maintained at T_{er} if the q-axis stator current i_{mq} can track its rated value i_{mqr} and i_{md} is kept at 0.

The brief overall control approach is shown in Figure 3. The control approach consists three controllers: two stator current controllers and a rotation speed controller.

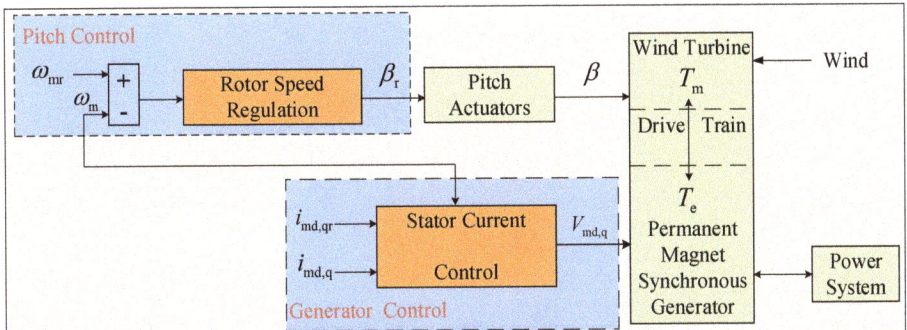

Figure 3. Brief overall control structure of the PMSG-WT.

3. Perturbation Observer-Based Nonlinear Adaptive Controller Design

In this section, the design of NAC for the PMSG-WT based on the feedback linearization will be presented. The NAC based on perturbation estimation proposed in [29] will be used. First, a nonlinear system is transformed as interacted subsystems through input/output linearization. Secondly, in each subsystem, uncertainties, nonlinearities and interaction among subsystems are contained in a defined perturbation term. The perturbation term is estimated via a designed observer. The estimated perturbation is used for compensating the real perturbation and obtaining adaptive linearizing control of the original nonlinear system.

3.1. NAC Design of WT

3.1.1. Input/Output Linearization

Input/output linearization of WT speed dynamics in system Equation (6) can be represented as

$$y_1^{(2)} = \frac{1}{J_{tot}}(\dot{T}_m - \dot{T}_e) \tag{8}$$

As the electromagnetic torque has much faster response than the mechanical torque, from the perspective of control of WT, $\dot{T}_e \simeq 0$. Equation (8) can be expressed as

$$\begin{aligned} y_1^{(2)} &= \frac{1}{J_{tot}}\dot{T}_m \\ &= F_1(x) + B1(x)u_1 \end{aligned} \tag{9}$$

where

$$\begin{aligned} F_1(x) &= A[-\frac{C_p}{\omega_m} - \frac{RV}{F^2}E]\frac{d\omega_m}{dt} \\ &\quad - \frac{AE\beta}{\tau_\beta}[-0.088e^{-12.5\tau} - \frac{0.08V^2}{F} + \frac{0.105\beta^2}{(1+\beta^3)^2}]\frac{d\beta}{dt} \end{aligned} \tag{10}$$

$$B1(x) = \frac{AE\beta}{\tau_\beta}[-0.088e^{-12.5\tau} - \frac{0.08V^2}{F} + \frac{0.105\beta^2}{(1+\beta^3)^2}] \tag{11}$$

where

$$\begin{aligned} A &= \frac{\rho\pi R^2 V^3}{2\omega_m} \\ E &= (39.27 - 319\tau + 1.1\beta)e^{-12.5\tau} \\ F &= \omega_m R + 0.08\beta V \\ \tau &= \frac{1}{\lambda + 0.08\beta} - \frac{0.035}{\beta^3 + 1} \end{aligned} \tag{12}$$

Please note that $\frac{dV}{dt}$ is not included in the FLC design, which cannot be directly measured.

As $\det[B1(x)] = \frac{AE\beta}{\tau_\beta}[-0.088e^{-12.5\tau} - \frac{0.08V^2}{F} + \frac{0.105\beta^2}{(1+\beta^3)^2}] \neq 0$ when $V \neq 0$ and $\beta \neq 0$, i.e., $B1(x)$ is nonsingular for all nominal operation points. Therefore, the FLC is expressed as

$$u_1 = B1(x)^{-1}(-F_1(x) + v_1) \quad (13)$$

And the nonlinear system is linearized as

$$y_1^{(2)} = v_1 \quad (14)$$

$$v_1 = \ddot{y}_{1r} + k_{11}\dot{e}_1 + k_{12}e_1 \quad (15)$$

where v_1 is input of linear systems, k_{11} and k_{12} are gains of linear controller, y_{1r} is the output reference, and $e_1 = y_{1r} - y_1$ as tracking error. The error dynamic is

$$\ddot{e}_1 + k_{11}\dot{e}_1 + k_{12}e_1 = 0 \quad (16)$$

3.1.2. Definition of Perturbation and State

For this subsystem, a perturbation term including all subsystem uncertainties, nonlinearities and interactions among subsystems is defined.

Define perturbation term $\Psi_1(x)$ as:

$$S_1: \begin{cases} \Psi_1(x) = F_1(x) + (B1(x) - B1(0))u_1 \\ B1(0) = \frac{AE\beta}{\tau_\beta}[-0.088e^{-12.5\tau} - \frac{0.08V^2}{F} + \frac{0.105\beta^2}{(1+\beta^3)^2}] \end{cases} \quad (17)$$

where $B1(0)$ is nominal value of $B1(x)$.

Defining the state vectors as $z_{11} = y_1, z_{12} = y_1^{(1)}, z_{13} = \Psi_1$, and control variable as $u_1 = \beta_r$. The dynamic equation of the subsystem S_1 becomes as

$$S_1: \begin{cases} z_{11} = y_1 \\ \dot{z}_{11} = z_{12} \\ \dot{z}_{12} = \Psi_1(x) + B1(0)u_1 \\ \dot{z}_{13} = \dot{\Psi}_1(x) \end{cases} \quad (18)$$

For subsystem S_1, several types of perturbation observers, e.g., linear Luenberger observer, sliding mode observer and high-gain observer, have been proposed [19,29,36]. High-gain observers proposed in [29] are used to estimate states and perturbations in this paper.

3.1.3. Design of States and Perturbation Observer

When the system output y_1 is available, a third-order SPO is employed for estimations of states and perturbation of the subsystem, which is designed as

$$S1: \begin{cases} \dot{\hat{z}}_{11} = \hat{z}_{12} + l_{11}(z_{11} - \hat{z}_{11}) \\ \dot{\hat{z}}_{12} = \hat{z}_{13} + l_{12}(z_{11} - \hat{z}_{11}) + B1(0)u_1 \\ \dot{\hat{z}}_{13} = l_{13}(z_{11} - \hat{z}_{11}), \end{cases} \quad (19)$$

where $\hat{z}_{11}, \hat{z}_{12}$ and $\hat{z}_{13}$ are the estimations of z_{11}, z_{12} and z_{13}, respectively, and l_{11}, l_{12} and l_{13} are gains of the observers, which are designed as

$$l_{ij} = \frac{\alpha_{ij}}{\epsilon_i^j} \quad (20)$$

where $i = 1, 2, 3; j = 1, \cdots, r_i + 1$, ϵ_i is a scalar chosen to be within $(0,1)$ for representing times of the time-dynamics between the real system and the observer, and parameters α_{ij} are chosen so that the roots of

$$s^{r_i+1} + \alpha_{i1}s^{r_i} + \cdots + \alpha_{ir_i}s + \alpha_{i(r_i+1)} = 0 \quad (21)$$

are in the open left-half complex plane.

3.1.4. Design of Nonlinear Adaptive Controller

The estimated perturbation is used for compensating the real perturbation, and control laws of subsystem S_1 can be obtained as follows:

$$u_1 = B1(0)^{-1}(-\hat{z}_{13} + v_1) \quad (22)$$

where v_1 is defined as

$$v_1 = \ddot{z}_{11r} + k_{12}(z_{11r} - \hat{z}_{11}) + k_{11}(\dot{z}_{11r} - \hat{z}_{12}) \quad (23)$$

3.2. NAC Design of PMSG

3.2.1. Input/Output Linearization

Input/output linearization of Equation (6) is represented as

$$\begin{bmatrix} y_2^{(1)} \\ y_3^{(1)} \end{bmatrix} = \begin{bmatrix} F_2(x) \\ F_3(x) \end{bmatrix} + B2(x) \begin{bmatrix} u_2 \\ u_3 \end{bmatrix} \quad (24)$$

where

$$F_2(x) = \frac{1}{L_{md}}(-i_{md}R_s + \omega_e L_{mq}i_{mq}) \quad (25)$$

$$F_3(x) = -\frac{R_s}{L_{mq}}i_{mq} - \frac{1}{L_{mq}}\omega_e(L_{md}i_{md} + K_e) \quad (26)$$

$$\quad (27)$$

$$B2(x) = \begin{bmatrix} B_2(x) \\ B_3(x) \end{bmatrix} = \begin{bmatrix} \frac{1}{L_{md}} & 0 \\ 0 & \frac{1}{L_{mq}} \end{bmatrix} \quad (28)$$

As $\det[B2(x)] = \frac{1}{L_{md}L_{mq}} \neq 0$, i.e., $B(x)$ is nonsingular for all nominal operation points. Therefore, the FLC controller is represented as

$$\begin{pmatrix} u_2 \\ u_3 \end{pmatrix} = B2(x)^{-1} \begin{pmatrix} -F_2(x) + v_2 \\ -F_3(x) + v_3 \end{pmatrix} \tag{29}$$

$$B2(x)^{-1} = \begin{bmatrix} L_{md} & 0 \\ 0 & L_{mq} \end{bmatrix} \tag{30}$$

And the nonlinear system is linearized as

$$\begin{bmatrix} y_2^{(1)} \\ y_3^{(1)} \end{bmatrix} = \begin{bmatrix} v_2 \\ v_3 \end{bmatrix} \tag{31}$$

where

$$v_2 = \dot{y}_{2r} + k_{21} e_2 \tag{32}$$
$$v_3 = \dot{y}_{3r} + k_{31} e_3 \tag{33}$$

where v_2 and v_3 are inputs of linear systems, k_{21} and k_{31} are gains of linear controller, y_{2r} and y_{3r} the output references. Define $e_2 = y_{2r} - y_2$ and $e_3 = y_{3r} - y_3$ as tracking errors, the error dynamics are

$$\dot{e}_2 + k_{21} e_2 = 0 \tag{34}$$
$$\dot{e}_3 + k_{31} e_3 = 0 \tag{35}$$

3.2.2. Definition of Perturbation and State

Define perturbation terms $\Psi_{2,3}(x)$ as:

$$S_2 : \begin{cases} \Psi_2(x) = F_2(x) + (B_2(x) - B_2(0)) \begin{bmatrix} u_2 \\ u_3 \end{bmatrix}, \\ B_2(0) = \begin{bmatrix} \frac{1}{L_{md0}} & 0 \end{bmatrix} \end{cases}$$

$$S_3 : \begin{cases} \Psi_3(x) = F_3(x) + (B_3(x) - B_3(0)) \begin{bmatrix} u_2 \\ u_3 \end{bmatrix} \\ B_3(0) = \begin{bmatrix} 0 & \frac{1}{L_{mq0}} \end{bmatrix} \end{cases} \tag{36}$$

where L_{md0} and L_{mq0}, $B_2(0)$ and $B_3(0)$ are nominal values of L_{md}, L_{mq}, $B_2(x)$ and $B_3(x)$, respectively.

Defining the state vectors as $z_{21} = y_2, z_{22} = \Psi_2$ and $z_{31} = y_3, z_{32} = \Psi_3$, and control variables as $u_2 = V_{md}$ and $u_3 = V_{mq}$. The dynamic equations of the two subsystems S_2 and S_3 become as

$$S_2 : \begin{cases} z_{21} = y_2 \\ \dot{z}_{21} = \Psi_2(x) + B_2(0) \begin{bmatrix} u_2 \\ u_3 \end{bmatrix}, \\ \dot{z}_{22} = \dot{\Psi}_2(x) \end{cases}$$

$$S_3 : \begin{cases} z_{31} = y_3 \\ \dot{z}_{31} = \Psi_3(x) + B_3(0) \begin{bmatrix} u_2 \\ u_3 \end{bmatrix}, \\ \dot{z}_{32} = \dot{\Psi}_3(x) \end{cases} \tag{37}$$

3.2.3. Design of Perturbation Observer

When the system outputs $y_{2,3}$ are available, two second-order POs are designed for the estimations of states and perturbation for the subsystems

$$S2: \begin{cases} \dot{\hat{z}}_{21} = \hat{z}_{21} + l_{21}(z_{21} - \hat{z}_{21}) + B_2(0)u_2 \\ \dot{\hat{z}}_{22} = l_{22}(z_{21} - \hat{z}_{21}), \end{cases} \quad (38)$$

$$S3: \begin{cases} \dot{\hat{z}}_{31} = \hat{z}_{31} + l_{31}(z_{31} - \hat{z}_{31}) + B_3(0)u_3 \\ \dot{\hat{z}}_{32} = l_{32}(z_{31} - \hat{z}_{31}), \end{cases} \quad (39)$$

where $\hat{z}_{21}$, $\hat{z}_{22}$, $\hat{z}_{31}$, and $\hat{z}_{32}$ are the estimations of z_{21}, z_{22}, z_{31} and z_{32}, respectively, and l_{21}, l_{22}, l_{31} and l_{32} are gains of the observers. They are designed similarly to Equation (20).

Remark 1. *It should be mentioned that during the design procedure, ϵ_i used in POs Equations (38) and (39) are required to be some relatively small positive constants only, and the performance of POs is not very sensitive to the observer gains, which are determined based on the upper bound of the derivative of perturbation.*

3.2.4. Design of Nonlinear Adaptive Controller

The estimated perturbations are used for compensating the real perturbation, and control laws of subsystems S_2 and S_3 can be obtained as follows:

$$\begin{bmatrix} u_2 \\ u_3 \end{bmatrix} = B2(0)^{-1}\left[\begin{bmatrix} -\hat{z}_{22} \\ -\hat{z}_{32} \end{bmatrix} + \begin{bmatrix} v_2 \\ v_3 \end{bmatrix} \right] \quad (40)$$

where $v_{2,3}$ is defined as

$$\begin{cases} v_2 = k_{21}(z_{21r} - \hat{z}_{21}) + \dot{z}_{21r} \\ v_3 = k_{31}(z_{31r} - \hat{z}_{31}) + \dot{z}_{31r} \end{cases} \quad (41)$$

The final control law represented by currents and inductances, are expressed as follows:

$$\begin{cases} u_2 = L_{md0}[k_{21}(i_{mdr} - i_{md}) + \dot{i}_{mdr} - \hat{\Psi}_2] \\ u_3 = L_{mq0}[k_{31}(i_{mqr} - i_{mq}) + \dot{i}_{mqr} - \hat{\Psi}_3] \end{cases} \quad (42)$$

Please note that only the nominal values of L_{md0}, L_{mq0}, and measurements of i_{md} and i_{mq} are required in the NAC design.

To clearly illustrate its principle, Figure 4 shows the block diagram of the NAC.

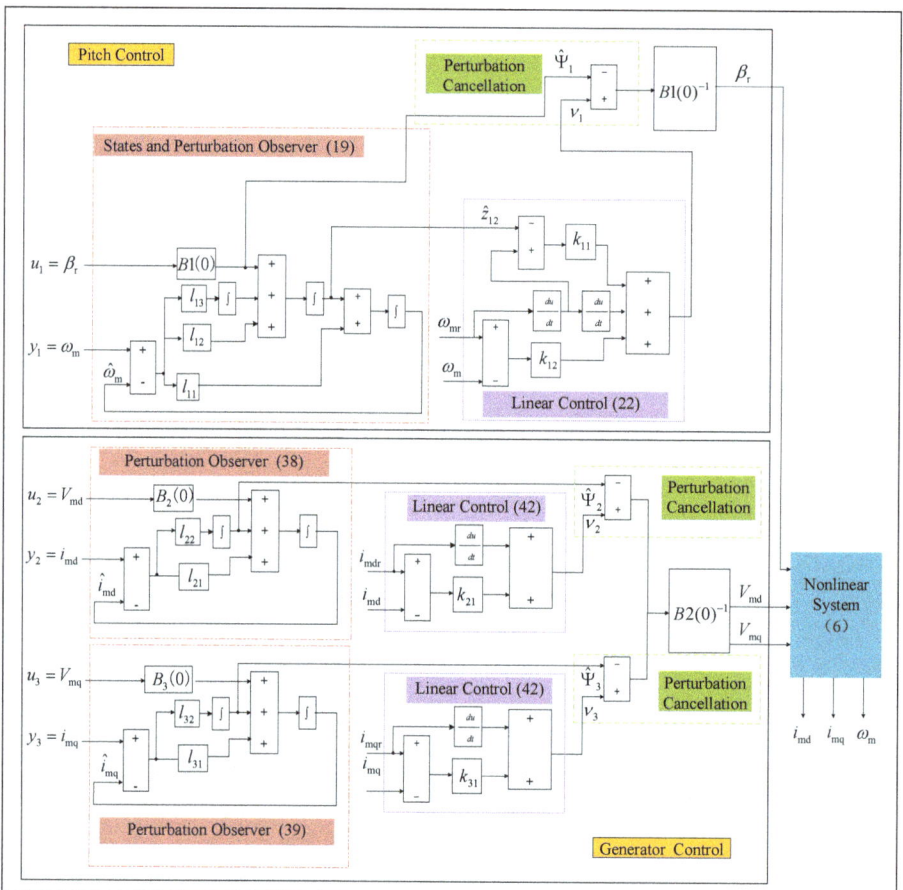

Figure 4. Block diagram of nonlinear adaptive controller.

The following assumptions are made in [19,21,36–39].

Assumption 1. *Input gain $B(x)$ and its derivative are bounded by $0 < M_1 \leq B(x) \leq M_2$, $|\dot{B}(x)| \leq M_3$, where M_i, $i = 1,2,3$ are finite constants [for convenience we assume that $B(x) > 0$]. $B(0)$ is chosen to satisfy: $|B(x)/B(0) - 1| \leq \theta < 1$, where θ is a positive constant. The control u is assumed to be bounded but big enough for the purpose of perturbation cancellation.*

Assumption 2. *The perturbation $\Psi_i(x,t)$ and its derivative $\dot{\Psi}_i(x,t)$ are locally Lipschitz in their arguments and bounded over the domain of interest.*

3.2.5. Stability Analysis of Closed-Loop System

This subsection analyzes the stability of the closed-loop system equipped with the NAC designed in the previous section.

At first, both the estimation error system and the tracking error system are obtained. On one hand, by defining estimation errors $\varepsilon_{21} = z_{21} - \hat{z}_{21}$, $\varepsilon_{22} = z_{22} - \hat{z}_{22}$, $\varepsilon_{31} = z_{31} - \hat{z}_{31}$, $\varepsilon_{32} = z_{32} - \hat{z}_{32}$, subtracting Equation (38) from Equation (37) and subtracting Equation (39) from Equation (37), the following estimation error system yields:

$$\dot{\varepsilon}_i = A_i \varepsilon_i + \eta_i \tag{43}$$

where

$$\varepsilon_i = \begin{bmatrix} \varepsilon_{21} \\ \varepsilon_{22} \\ \varepsilon_{31} \\ \varepsilon_{32} \end{bmatrix}, \quad \eta_i = \begin{bmatrix} 0 \\ \Psi_2 \\ 0 \\ \Psi_3 \end{bmatrix},$$

$$A_i = \begin{bmatrix} -l_{21} & 1 & 0 & 0 \\ -l_{22} & 0 & 0 & 0 \\ 0 & 0 & -l_{31} & 1 \\ 0 & 0 & -l_{32} & 0 \end{bmatrix} \tag{44}$$

On the other hand, define the tracking errors as $e_{21} = y_{2r} - z_{21}$ and $e_{31} = y_{3r} - z_{31}$. It follows from Equations (24), (26), (36), (40) and (41) that

$$\begin{bmatrix} \dot{e}_{21} \\ \dot{e}_{31} \end{bmatrix} = - \begin{bmatrix} k_{21}(e_{21} + \varepsilon_{21}) + \varepsilon_{22} \\ k_{31}(e_{31} + \varepsilon_{31}) + \varepsilon_{32} \end{bmatrix} \tag{45}$$

Thus, the tracking error system can be summarized as

$$\dot{e}_i = M_i e_i + \vartheta_i \tag{46}$$

where

$$e_i = \begin{bmatrix} e_{21} \\ e_{22} \end{bmatrix}, \quad \vartheta_i = \begin{bmatrix} -\xi_1 \\ -\xi_2 \end{bmatrix},$$

$$M_i = \begin{bmatrix} -k_{21} & 0 \\ 0 & -k_{31} \end{bmatrix} \tag{47}$$

with $\xi_1 = \varepsilon_{22} + k_{21}\varepsilon_{21}$ and $\xi_2 = \varepsilon_{32} + k_{31}\varepsilon_{31}$ being the lumped estimation error.

The stability analysis of the closed-loop control system is transformed into globally uniformly ultimately bounded summarized.

Theorem 1. *Consider the PMSM system Equation (24) equipped the proposed NAC Equation (42) with two POs Equations (38) and (39). If the real perturbation $\Psi_i(x,t)$ defined in Equation (36) satisfies*

$$\|\Psi_i(x,t)\| \leq \gamma_1 \tag{48}$$

then both the estimation error system Equation (43) and the tracking error system Equation (46) are, i.e.,

$$\|\varepsilon_i(t)\| \leq 2\gamma_1 \|P_1\|, \|e_i(t)\| \leq 4\gamma_1 \|K_i\| \|P_1\| \|P_2\|, \forall t \geq T \tag{49}$$

where P_i, $i = 1, 2$ are respectively the feasible solutions of Riccati equations $A_i^T P_1 + P_1 A_i = -I$ and $M_i^T P_2 + P_2 M_i = -I$; and $\|K_i\|$ is a constant related to k_{11}, k_{21} and k_{22}.

Proof. For the estimation error system Equation (43), consider the following Lyapunov function:

$$V_{i1}(\varepsilon_i) = \varepsilon_i^T P_1 \varepsilon_i \tag{50}$$

The high gains of POs Equations (38) and (39) are determined by requiring Equation (21) holds, which means A_i is Hurwitz. One can find a feasible positive definite solution, P_1, of Riccati equation $A_i^T P_1 + P_1 A_i = -I$. Calculating the derivative of $V_{i1}(\varepsilon_i)$ along the solution of system Equation (43) and using Equation (48) to yield

$$\begin{aligned}
\dot{V}_{i1}(\varepsilon_i) &= \varepsilon_i^T(A_i^T P_1 + P_1 A_i)\varepsilon_i + \eta_i^T P_1 \varepsilon_i + \varepsilon_i^T P_1 \eta_i \\
&\leq -\|\varepsilon_i\|^2 + 2\|\varepsilon_i\| \cdot \|\eta_i\| \cdot \|P_1\| \\
&\leq -\|\varepsilon_i\|(\|\varepsilon_i\| - 2\gamma_1 \|P_1\|)
\end{aligned} \quad (51)$$

Then $\dot{V}_{i1}(\varepsilon_i) \leq 0$ when $\|\varepsilon_i\| \geq 2\gamma_1 \|P_1\|$. Thus, there exists $T_1 > 0$, which can lead to

$$\|\varepsilon_i(t)\| \leq \gamma_2 = 2\gamma_1 \|P_1\|, \forall t \geq T_1 \quad (52)$$

For tracking error system Equation (46), one can find that $\|\vartheta_i\| \leq \|K_i\|\gamma_2$ with $\|K_i\|$ based on $\|\varepsilon_i(t)\| \leq \gamma_2$. Consider the Lyapunov function $V_{i2}(e_i) = e_i^T P_2 e_i$. Similarly, one can prove that there exists an instant, $\bar{T}_1$, the following holds

$$\|e_i(t)\| \leq 2\|K_i\|\gamma_2 \|P_2\| \leq 4\gamma_1 \|K_i\| \|P_1\| \|P_2\|, \forall t \geq \bar{T}_1 \quad (53)$$

Using Equations (52) and (53) and setting $T = \max\{T_1, \bar{T}_1\}$ lead to Equation (49).

Moreover, if $\Psi_i(x,t)$ and $\dot{\Psi}_i(x,t)$ are locally Lipschitz in their arguments, it will guarantee the exponential convergence of the observation error [19] and closed-loop tracking error into

$$\lim_{t \to \infty} \varepsilon_i(t) = 0 \quad \text{and} \quad \lim_{t \to \infty} e_i(t) = 0 \quad (54)$$

After the states i_d and i_q and their derivatives are stable that controlled by NAC. The parameter variation is considered in the error system in Equations (43) and (46), and the error system is proved as converged to zero in Equation (54). This guarantees that the estimated perturbations track the extended states defined in Equation (36), which includes the uncertainties affected by the parameter variations and disturbances, and compensates for the control input in Equation (40). Then the linearized subsystems in Equation (37) are independent of the parameters and disturbances. □

Remark 2. *The perturbation and its derivative are assumed to locally bounded as described in* **Assumption 2***. The existence of these bounds can be shown in the following analysis. The perturbation and its derivative can be represented as*

$$\begin{aligned}
\Psi_2 &= F_2(x) + \frac{B_2(x)-B_2(0)}{B_2(x)}[k_{21}(z_{21r}-z_{21}) + z_{22} - \hat{z}_{22}] \\
&= F_2(x) + \frac{B_2(x)-B_2(0)}{B_2(x)}(k_{21}e_{21} + \varepsilon_{22}) \\
\dot{\Psi}_2 &= \dot{F}_2(x) + \frac{B_2(x)-B_2(0)}{B_2(0)}(-\dot{\Psi}_2 + k_{21}\dot{e}_{21} - \dot{\varepsilon}_{22}) \\
&= \dot{F}_2(x) + \frac{B_2(x)-B_2(0)}{B_2(0)}(k_{21}\dot{e}_{21} + l_{22}\varepsilon_{21}) \\
\Psi_3 &= F_3(x) + \frac{B_3(x)-B_3(0)}{B_3(x)}[k_{31}(z_{31r}-z_{31}) + z_{32} - \hat{z}_{32}] \\
&= F_3(x) + \frac{B_3(x)-B_3(0)}{B_3(x)}(k_{31}e_{31} + \varepsilon_{32}) \\
\dot{\Psi}_3 &= \dot{F}_3(x) + \frac{B_3(x)-B_3(0)}{B_3(0)}(-\dot{\Psi}_3 + k_{31}\dot{e}_{31} - \dot{\varepsilon}_{32}) \\
&= \dot{F}_3(x) + \frac{B_3(x)-B_3(0)}{B_3(0)}(k_{31}\dot{e}_{31} + l_{32}\varepsilon_{31})
\end{aligned}$$

Considering **Assumption 1**, we have

$$|\Psi_2| \leq \frac{1}{1-\theta_2}|F_2(x)| + \frac{\theta_2}{1+\theta_2}(\|k_{21}\|\|e_{21}\| + |\varepsilon_{22}|)$$

$$|\dot{\Psi}_2| \leq |\dot{F}_2(x)| + |B_2(x)||u_2| + \theta_2(\|k_{21}\|\|\dot{e}_{21}\| + l_{22}|\varepsilon_{21}|)$$

$$|\Psi_3| \leq \frac{1}{1-\theta_3}|F_3(x)| + \frac{\theta_3}{1+\theta_3}(\|k_{31}\|\|e_{31}\| + |\varepsilon_{32}|)$$

$$|\dot{\Psi}_3| \leq |\dot{F}_3(x)| + |B_3(x)||u_3| + \theta_3(\|k_{31}\|\|\dot{e}_{31}\| + l_{32}|\varepsilon_{31}|)$$

From the above equations, with consideration of the perturbation assumed as a smooth function of time, it can be concluded that the bound of perturbation and its derivative exist.

As a result, with both the **Assumptions 1** and **2**, the effectiveness of such perturbation observer-based control can be guaranteed.

4. Simulation Results

To verify the effectiveness of the proposed NAC, simulations studies (Matlab/Simulink) have been carried out by comparing with the VC, GSPI and FLC. In this paper, a 2 MW PMSG-WT given in [35] is investigated. The parameters of the PMSG-WT system are listed in Table 1. In this paper, the mechanical rotation speed reference is $\omega_{mr} = 2.2489$ rad/s. The reference of d-axis stator current is $i_{mdr} = 0$ A. The rated electromagnetic torque reference is $T_{er} = 889326.7$ Nm. According to Equation (7), the q-axis stator current reference is $i_{mqr} = 593.3789$ A.

Table 1. Parameters of PMSG-WT for simulation studies.

Parameters	Values	Units
Air density ρ	1.205	kg/m^3
Rated wind speed V_r	12	m/s
Blade radius R	39	m
Actuator time constant τ_β	1	s
pitch angle rate β_{rate}	±10	degree/s
Rated output power P_r	2	MW
Stator resistance R_s	50	μΩ
d-axis inductance L_d	5.5	mH
q-axis inductance L_q	3.75	mH
Number of pole pairs p	11	
Field flux K_e	136.25	V·s/rad
Total inertia J_{tot}	10,000	kg·m^2

Parameters of NACs for subsystems S_1, S_2, and S_3 are designed based on pole-placement and listed in Table 2. Please note that the controller parameters of the FLC are the same as that of the NAC for all three subsystems, and the FLC requires exact system parameters and full state measurements except $\frac{dV}{dt}$.

Table 2. Parameters of Pitch control approach for simulation studies.

Parameters of the NAC Equation (42)	
Gains of observer Equation (19)	$\alpha_{11} = 50, \alpha_{12} = 1.875 \times 10^3, \alpha_{13} = 1.5625 \times 10^4, \epsilon_1 = 0.02$
Gains of observer Equation (38)	$\alpha_{21} = 4 \times 10^2, \alpha_{22} = 4 \times 10^4, \epsilon_2 = 0.01$
Gains of observer Equation (39)	$\alpha_{31} = 4 \times 10^2, \alpha_{32} = 4 \times 10^4, \epsilon_3 = 0.01$
Gains of linear controller Equation (41)	$k_{11} = 40, k_{12} = 4 \times 10^2, k_{21} = 1.6 \times 10^2, k_{31} = 1.6 \times 10^2$

4.1. Ramp Wind

Figures 5 and 6 show the responses of the PMSG-WT to ramp wind. Wind speed is shown in Figure 5a. As shown in Figure 5b,c, the proposed NAC provides the smallest tracking error of the mechanical rotation speed ω_m, compared with the VC, GSPI and FLC. The VC has the biggest tracking error and requires the longest recovery time. It can be explained that the VC is adjusted for a specific operation point of the system and cannot ensure provision of a satisfactory dynamic performance for time-varying operation points. Although the FLC can provide a high tracking performance, the tracking error of ω_m still exists. It is because that the FLC requires full state measurements, but the $\frac{dV}{dt}$ in Equation (9) is unknown in the FLC design. The GSPI also achieves better performance than the VC. This is because the GSPI can schedule PI gains frequently under time-varying wind speeds. However, it increases the burden of the controller.

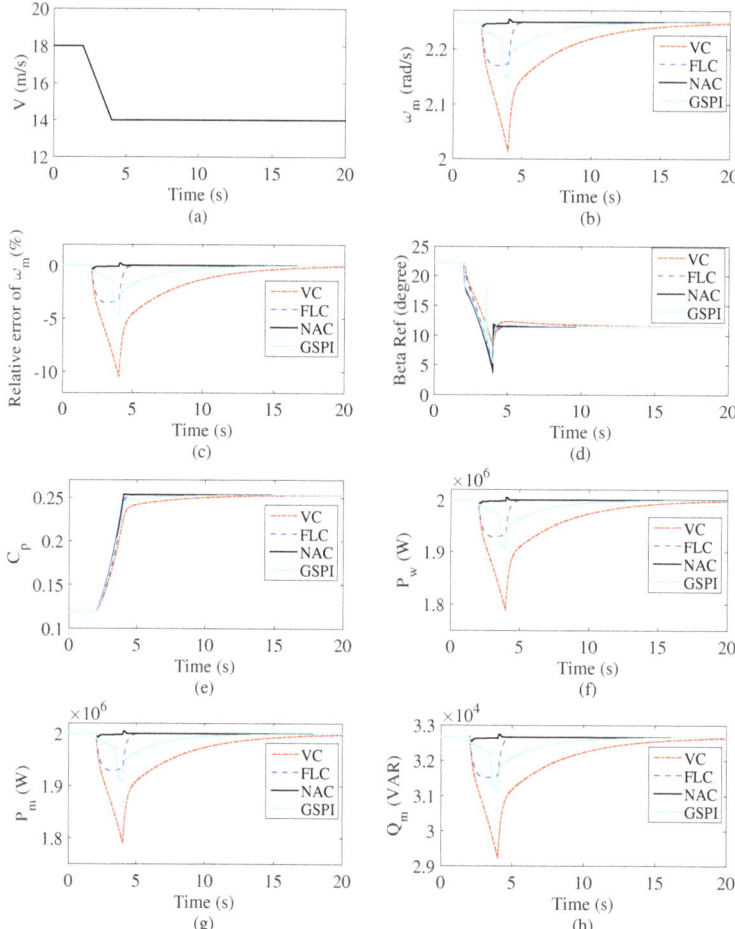

Figure 5. Responses of the PMSG-WT to ramp wind speed. (**a**) Wind speed V. (**b**) Mechanical rotation speed ω_m. (**c**) Relative error of ω_m. (**d**) Required pitch angle. (**e**) Power coefficient C_p. (**f**) Mechanical power P_w. (**g**) Active generating power P_m. (**h**) Reactive generating power Q_m.

To keep the extracted wind power at the rated power, the required pitch angle β_r should change with the varying wind speed, as shown in Figure 5d. In Figure 5e,f, to maintain the extracted wind power around its rated value, the power coefficient C_p increases when wind speed decreases. The extracted wind power can be maintained around its rated value under the NAC even when wind speed varies, which the VC, GSPI and FLC cannot provide. The active generating power P_m and reactive generating power Q_m of the PMSG-WT are shown in Figure 5g,h, respectively.

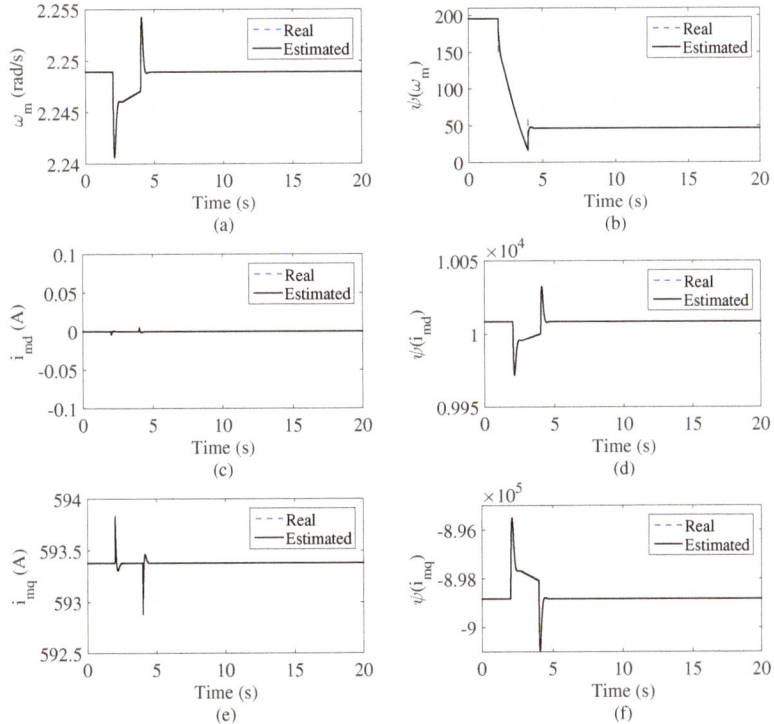

Figure 6. Estimations of states and perturbations. (**a**) Estimation of mechanical rotation speed ω_m. (**b**) Estimation of perturbation term Ψ_1. (**c**) Estimation of i_{md}. (**d**) Estimation of perturbation term Ψ_2. (**e**) Estimation of i_{mq}. (**f**) Estimation of perturbation term Ψ_3.

In the previous section, it mentions that the proposed NAC can estimate the defined perturbation terms Equations (17) and (36) via the designed observers Equations (19), (38) and (39) to compensate the real perturbation. It can be seen from Figure 6 that both the states and perturbations can be well estimated by the designed SPO.

4.2. Random Wind

Figures 7 and 8 show the responses of the PMSG-WT to random wind. Figure 7a shows time-varying wind speed. It can be seen from Figure 7b,c that the VC, GSPI and FLC cannot provide high tracking performance of the mechanical rotation speed ω_m under time-varying wind speed. However, the GSPI achieve better tracking performance than the FLC under random wind speeds. The NAC always keeps mechanical rotation speed ω_m around its rated value. To limit the extracted wind power, the power coefficient C_p varies with time-varying wind speed, shown in Figure 7d. During the whole operating period, the NAC can always keep consistent responses of P_m and Q_m shown in Figure 7e,f. The performances of the VC, GSPI and FLC are all affected by the time-varying wind speed. Figure 8 shows the designed observers can provide satisfactory estimations for the states and perturbations.

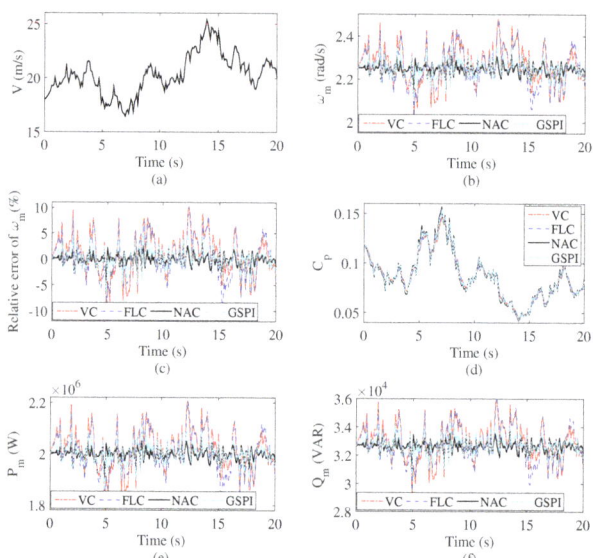

Figure 7. Responses of the PMSG-WT to random wind speed. (**a**) Wind speed V. (**b**) Mechanical rotation speed ω_m. (**c**) Relative error of ω_m. (**d**) Power coefficient C_p. (**e**) Active generating power P_m. (**f**) Reactive generating power Q_m.

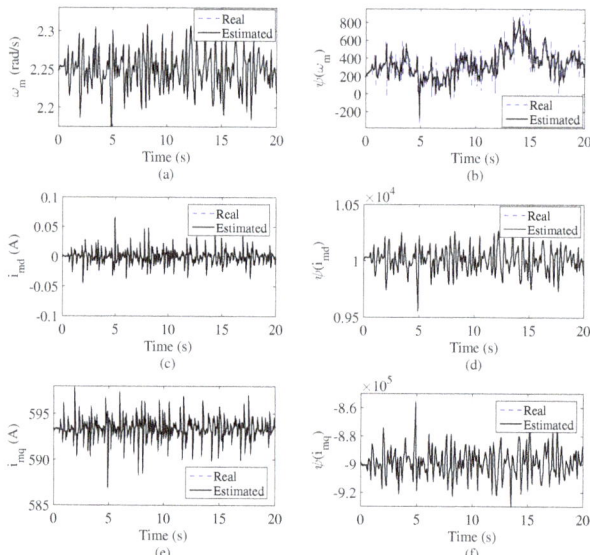

Figure 8. Estimations of states and perturbations. (**a**) Estimation of mechanical rotation speed ω_m. (**b**) Estimation of perturbation term Ψ_1. (**c**) Estimation of i_{md}. (**d**) Estimation of perturbation term Ψ_2. (**e**) Estimation of i_{mq}. (**f**) Estimation of perturbation term Ψ_3.

4.3. Robustness Against Parameter Uncertainty

For a practical PMSG-WT system, the operating temperature, manufacturing tolerance and magnetic saturation effect may result in the variation of system parameter values. The control performance of the VC, FLC and proposed NAC is tested under field flux variation. Please note that the wind speed is kept at 18 m/s. The variation of field flux K_e is shown in Figure 9a.

In Figure 9b, the proposed NAC can provide better tracking performance of the mechanical rotation speed ω_m, compared with the VC and FLC. The maximum relative error ($\frac{\omega_m - \omega_{mr}}{\omega_{mr}} \times 100\%$) reaches approximately 5% and 1% under the FLC and VC, respectively. The control performance of the VC and FLC are both affected by field flux variation. In Figure 9c,d, the responses of the required pitch angle β_r and power coefficient C_p are shown. The active generating power P_m and reactive generating power Q_m of the PMSG-WT are shown in Figure 9e,f, respectively. The active generation power P_m cannot be kept at its rated value under these three controllers, especially under the FLC. The reactive generating power Q_m is almost unaffected under the NAC.

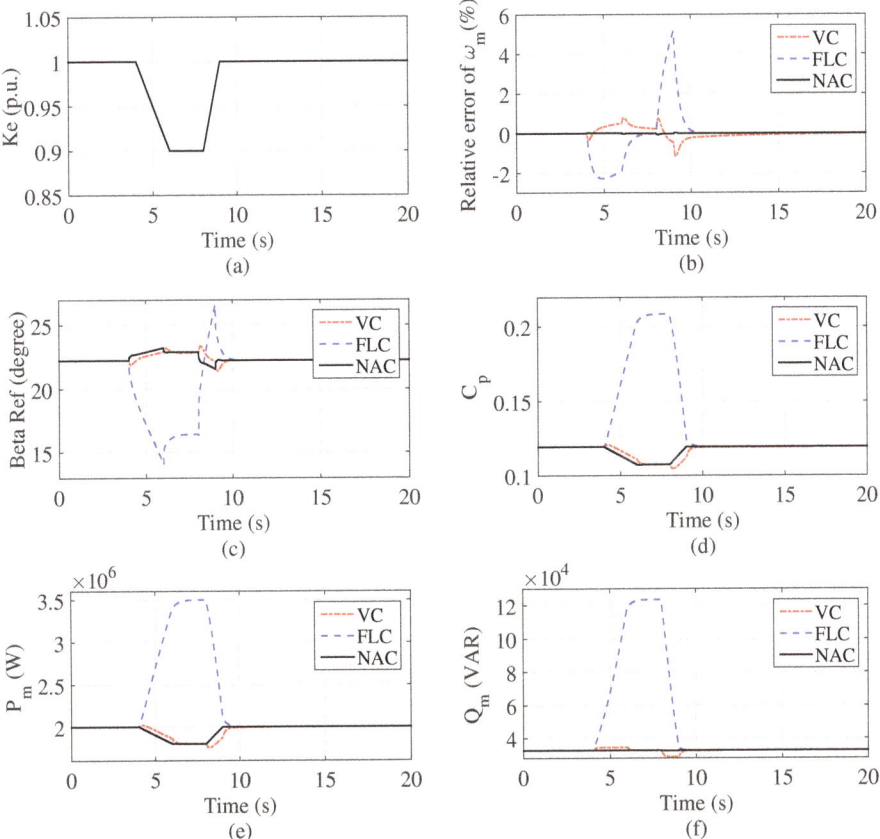

Figure 9. Response to field flux K_e variation under constant wind speed. (**a**) Variation of field flux K_e. (**b**) Relative error of mechanical rotation speed ω_m. (**c**) Required pitch angle. (**d**) Power coefficient C_p. (**e**) Active generating power P_m. (**f**) Reactive generating power Q_m.

In addition, Table 3 shows the control performance of these three controllers via integral of absolute error (IAE) in different simulation scenarios. Here, $IAE_x = \int_0^T |x - x^*|$. The reference value of the variable x is x^*. The simulation time T is set as 20 s. It can be seen from Table 3 that in first and second simulation scenarios, the IAE_{i_d} and IAE_{i_q} are both almost around 0 A.s under these three controllers. Compared with the VC and FLC, the IAE_{ω_m} is smaller under the proposed NAC. In the field flux variation simulation scenario, the NAC can provide much smaller IAE_{i_q} and IAE_{ω_m} than those achieved by the VC and FLC. Compare with the VC, the FLC is more significantly affected by field flux variation.

The proposed NAC can always provide a satisfactory performance. This is because the proposed NAC can estimate all uncertainties without knowing detailed system model. Therefore, it has better robustness than the FLC, which requires accurate system parameters. Meanwhile, the control performance of the VC is affected under parameter variations [29].

Table 3. IAE indices of different controllers in different scenarios.

Simulation Scenarios	Variables	Controllers		
		VC	FLC	NAC
Ramp wind speed	IAE_{ω_m} (rad)	0.817	0.1555	1.397×10^{-5}
	IAE_{i_d} (A·s)	9.603×10^{-15}	1.025×10^{-13}	3.076×10^{-5}
	IAE_{i_q} (A·s)	8.634×10^{-13}	6.972×10^{-12}	2.752×10^{-3}
Random wind speed	IAE_{ω_m} (rad)	1.369	1.273	1.514×10^{-3}
	IAE_{i_d} (A·s)	9.98×10^{-15}	1.075×10^{-13}	6.75×10^{-3}
	IAE_{i_q} (A·s)	9.598×10^{-13}	7.154×10^{-12}	0.6171
Field flux variation	IAE_{ω_m} (rad)	0.0695	0.207	2.995×10^{-6}
	IAE_{i_d} (A·s)	9.526×10^{-15}	1.24×10^{-13}	1.852×10^{-5}
	IAE_{i_q} (A·s)	67.78	1957	0.04528

5. Conclusions

This paper has developed a nonlinear adaptive pitch controller for the PMSG-WT to limit the extracted power from time-varying wind in Region 3. In the proposed NAC, all time-varying and unknown dynamics of the PMSG-WT, e.g., nonlinearities, parameter uncertainties and disturbances, are included by defined perturbation terms, which are estimated by designed POs and SPO. The estimated perturbations are used to compensate the real perturbations for fully linearizing the PMSG-WT system. The proposed NAC has overcome the drawbacks of the FLC relying on the full system states and detailed nonlinear system model, the shortcoming of the VC designed based on a specific operating point, and the disadvantages of the GSPI scheduling PI gains frequently under time-varying wind speeds. Simulation studies are carried out for the comparison of the control performance achieved by the VC, FLC, GSPI and NAC under different scenarios. Compared with the FLC, GSPI and VC, the proposed NAC provides the best performance under different scenarios and achieves highest robustness against field flux variation. Wind speed sensorless control approach will be focused on in further work. The effective wind speed cannot be directly measured by anemometers, but it can be estimated through employing the WT itself as a wind speed measurement device [8].

Author Contributions: Conceptualization, J.C. and B.Y.; validation, J.C. and B.Y.; investigation, J.C. and W.D.; methodology, T.Y. and N.A.; writing—original draft preparation, J.C.; writing—review and editing, W.D. and B.Y.; visualization, H.S. and L.C.; supervision, H.S. and L.C.; project administration, J.C., B.Y. and T.Y.; funding acquisition, J.C., B.Y. and T.Y.

Funding: This research was funded by the Natural Science Foundation of the Jiangsu Higher Education Institutions of China (Grant No. 19KJB470036), National Natural Science Foundation of China grant number (Grant Nos. 51667010, 51777078), the Fundamental Research Funds for the Central Universities (Grant No. D2172920), the Key Projects of Basic Research and Applied Basic Research in Universities of Guangdong Province (Grant No. 2018KZDXM001), and the Science and Technology Projects of China Southern Power Grid (Grant No. GDKJXM20172831).

Conflicts of Interest: The authors declare no conflict of interest.

References

1. Aissaoui, A.G.; Tahour, A.; Essounbouli, A.; Nollet, F.; Abid M.; Chergui, M.I. A fuzzy-PI control to extract an optimal power from wind turbine. *Energy Convers. Manag.* **2013**, *65*, 688–696. [CrossRef]
2. Chen, J.; Yao, W.; Zhang, C.K.; Ren, Y.X.; Jiang, L. Design of robust MPPT controller for grid-connected PMSG-Based wind turbine via perturbation observation based nonlinear adaptive control. *Renew. Energy* **2019**, *134*, 478–495. [CrossRef]
3. Liao, S.W.; Yao, W.; Han, X.N.; Wen, J.Y.; Cheng, S.J. Chronological operation simulation framework for regional power system under high penetration of renewable energy using meteorological data. *Appl. Energy* **2017**, *203*, 816–828. [CrossRef]
4. Yang, B.; Jiang, L.; Yao, W.; Wu, Q.H. Nonlinear maximum power point tracking control and modal analysis of DFIG based wind turbine. *Int. J. Electr. Power Energy Syst.* **2016**, *74*, 429–436. [CrossRef]
5. Liu, J.; Yao, W.; Wen, J.Y.; Fang, J.K.; Jiang, L.; He, H.B.; Cheng, S.J. Impact of power grid strength and PLL parameters on stability of grid-connected DFIG wind farm. *IEEE Trans. Sustain. Energy* **2019**. [CrossRef]
6. Yang, B.; Zhang, X.S.; Yu, T.; Shu, H.C.; Fang, Z.H. Grouped grey wolf optimizer for maximum power point tracking of doubly-fed induction generator based wind turbine. *Energy Convers. Manag.* **2017**, *133*, 427–443. [CrossRef]
7. Abdullah, M.A.; Yatim, A.H.M.; Tan, C.W.; Saidur, R. A review of maximum power point tracking algorithms for wind energy systems. *Renew. Sustain. Energy Rev.* **2012**, *16*, 3220–3227. [CrossRef]
8. Kumar, A.; Stol, K. Simulating feedback linearization control of wind turbines using high-order models. *Wind Energy* **2009**, *13*, 419–432. [CrossRef]
9. Boukhezzar, B.; Siguerdidjane, H. Nonlinear Control of a variable-speed wind turbine using a two-mass model. *IEEE Trans. Energy Convers.* **2011**, *26*, 149–162. [CrossRef]
10. Van, T.L; Nguyem, T.H.; Lee, D.C. Advanced pitch angle control based on fuzzy logic for variable-speed wind turbine systems. *IEEE Trans. Energy Convers.* **2015**, *30*, 578–587. [CrossRef]
11. Boukhezzar, B.; Lupu, L.; Siguerdidjane, H.; Hand, M. Multivariable control strategy for variable speed, variable pitch wind turbines. *Renew. Energy* **2007**, *32*, 1273–1287. [CrossRef]
12. Connor, B.; Leithead, W.E.; Grimble, M. LQG control of a constant speed horizontal axis wind turbine. *Proc. IEEE CCA* **1994**, *1*, 251–252.
13. Bossanyi, E.A. The design of closed loop controllers for wind turbines. *Wind Energy* **2000**, *3*, 149–163. [CrossRef]
14. Akhmatov, V.; Knudsen, H.; Nielsen, A.H.; Pedersen, J.K.; Poulsen, N.J. Modelling and transient stability of large wind farms. *Int. J. Electr. Power Energy Syst.* **2003**, *25*, 123–144. [CrossRef]
15. Pao, L.; Johnson, K. A tutorial on the dynamics and control of wind turbines and wind farms. *Proc. ACC* **2009**. [CrossRef]
16. Mullane, A.; Lightbody, G.; Yacamini, R. Wind-turbine fault ride-through ehancement. *IEEE Trans. Power Syst.* **2005**, *20*, 1929–1937. [CrossRef]
17. Seol, J.Y.; Ha, I.J. Feedback-linearizing control of IPM motors considering magnetic saturation effect. *IEEE Trans. Power Electron.* **2005**, *20*, 416–424. [CrossRef]
18. Kim, K.H.; Jeung, Y.C.; Lee, D.C.; Kim, H.J. LVRT strategy of PMSG wind power systems based on feedback linearization. *IEEE Trans. Power Electron.* **2012**, *27*, 2376–2384. [CrossRef]
19. Jiang, L.; Wu, Q.H. Nonlinear adaptive control via sliding-mode state and perturbation observer. *IEEE Proc. Control Theory Appl.* **2002**, *149*, 269–277. [CrossRef]
20. Mauricio, J.M.; Leon, A.E.; Gomez-Exposito, A.; Solsona, J.A. An adaptive nonlinear controller for DFIM-based wind energy conversion systems. *IEEE Trans. Energy Convers.* **2008**, *23*, 1024–1035. [CrossRef]
21. Yang, B.; Yu, T.; Shu, H.C.; Dong, J.; Jiang, L. Robust sliding-mode control of wind energy conversion systems for optimal power extraction via nonlinear perturbation observers. *Appl. Energy* **2018**, *210*, 711–723. [CrossRef]
22. Noureldeen, O.; Hamdan, I. Design of robust intelligent protection technique for large-scale grid-connected wind farm. *Prot. Control Mod. Power Syst.* **2018**, *3*, 169–182. [CrossRef]
23. Shen, Y.; Yao, W.; Wen, J.Y.; He, H.B.; Jiang, L. Resilient wide-area damping control using GrHDP to tolerate communication failures. *IEEE Trans. Smart Grid* **2019**, *10*, 2547–2557. [CrossRef]

24. Han, B.; Zhou, L.W.; Yang, F; Xiang, Z. Individual pitch controller based on fuzzy logic control for wind turbine load mitigation. *IET Renew. Power Gener.* **2016**, *10*, 687–693. [CrossRef]
25. Liu, J.; Wen, J.Y.; Yao, W.; Long, Y. Solution to short-term frequency response of wind farms by using energy storage systems. *IET Renew. Power Gener.* **2016**, *10*, 669–678. [CrossRef]
26. Saravanakumar, R.; Jena, D. Validation of an integral sliding mode control for optimal control of a three blade variable speed variable pitch wind turbine. *Int. J. Electr. Power Energy Syst.* **2015**, *69*, 421–429. [CrossRef]
27. Saravanakumar, R.; Jena, D. Modified vector controlled DFIG wind energy system based on barrier function adaptive sliding mode control. *Prot. Control Mod. Power Syst.* **2019**, *4*, 34–41.
28. Lin, W.M.; Hong, C.M. A new Elman neural network-based control algorithm for adjustable-pitch variable-speed wind-energy conversion systems. *IEEE Trans. Power Electron.* **2011**, *26*, 473–481. [CrossRef]
29. Chen, J.; Jiang, L.; Yao, W.; Wu, Q.H. Perturbation estimation based nonlinear adaptive control of a full-rated converter wind-turbine for fault ride-through capability enhancement. *IEEE Trans. Power Syst.* **2014**, *29*, 2733–2743. [CrossRef]
30. Chen, J.; Yao, W.; Ren, Y.X.; Wang, R.T.; Zhang, L.H.; Jiang, L. Nonlinear adaptive speed control of a permanent magnet synchronous motor: A perturbation estimation approach. *Control Eng. Pract.* **2019**, *85*, 163–175. [CrossRef]
31. Yang, B.; Yu, T.; Shu, H.C.; Zhang, Y.M.; Chen, J.; Sang, Y.Y.; Jiang, L. Passivity-based sliding-mode control design for optimal power extraction of a PMSG based variable speed wind turbine. *Renew. Energy* **2018**, *119*, 577–589. [CrossRef]
32. Yang, B.; Yu, T.; Shu, H.C.; Zhu, D.N.; Zeng, F.; Sang, Y.Y.; Jiang, L. Perturbation observer based fractional-order PID control of photovoltaics inverters for solar energy harvesting via Yin-Yang-Pair optimization. *Energy Convers. Manag.* **2018**, *171*, 170–187. [CrossRef]
33. Ren, Y.X.; Li, L.Y.; Brindley, J.; Jiang, L. Nonlinear PI control for variable pitch wind turbine. *Control Eng. Pract.* **2016**, *50*, 84–94. [CrossRef]
34. Xia, Y.; Ahmed, K.H.; Williams, B.W. A new maximum power point tracking technique for permanent magnet synchronous generator based wind energy conversion system. *IEEE Trans. Power Electron.* **2011**, *26*, 3609–3620. [CrossRef]
35. Uehara, A.; Pratap, A.; Goya, T.; Senjyu, T.; Yona, A.; Urasaki, N.; Funabashi, T. A coordinated control method to smooth wind power fluctuations of a PMSG-based WECS. *IEEE Trans. Energy Convers.* **2011**, *26*, 550–558. [CrossRef]
36. Jiang, L.; Wu, Q.H.; Wen, J.Y. Decentralized nonlinear adaptive control for multimachine power systems via high-gain perturbation observer. *IEEE Trans. Circuits Syst. I Regul. Pap.* **2004**, *51*, 2052–2059. [CrossRef]
37. Youcef, K.; Wu, S. Input/output linearization using time delay control. *J. Dyn. Syst. Meas. Control* **1992**, *114*, 10–19. [CrossRef]
38. Yang, B.; Jiang, L.; Yao, W.; Wu, Q.H. Perturbation estimation based coordinated adaptive passive control for multimachine power systems. *Control Eng. Pract.* **2015**, *44*, 172–192. [CrossRef]
39. Wu, Q.H.; Jiang, L.; Wen, J.Y. Decentralized adaptive control of interconnected non-linear systems using high gain observer. *Int. J. Control* **2004**, *77*, 703–712. [CrossRef]

© 2019 by the authors. Licensee MDPI, Basel, Switzerland. This article is an open access article distributed under the terms and conditions of the Creative Commons Attribution (CC BY) license (http://creativecommons.org/licenses/by/4.0/).

Article

Global Maximum Power Point Tracking of PV Systems under Partial Shading Condition: A Transfer Reinforcement Learning Approach

Min Ding [1], Dong Lv [1], Chen Yang [1], Shi Li [1], Qi Fang [1], Bo Yang [2,*] and Xiaoshun Zhang [3]

1. State Grid Suzhou Power Supply Company, Suzhou 215004, China
2. Faculty of Electric Power Engineering, Kunming University of Science and Technology, Kunming 650500, China
3. College of Engineering, Shantou University, Shantou 515063, China
* Correspondence: yangbo_ac@outlook.com; Tel.: +86-183-1459-6103

Received: 11 June 2019; Accepted: 2 July 2019; Published: 9 July 2019

Abstract: This paper aims to introduce a novel maximum power point tracking (MPPT) strategy called transfer reinforcement learning (TRL), associated with space decomposition for Photovoltaic (PV) systems under partial shading conditions (PSC). The space decomposition is used for constructing a hierarchical searching space of the control variable, thus the ability of the global search of TRL can be effectively increased. In order to satisfy a real-time MPPT with an ultra-short control cycle, the knowledge transfer is introduced to dramatically accelerate the searching speed of TRL through transferring the optimal knowledge matrices of the previous optimization tasks to a new optimization task. Four case studies are conducted to investigate the advantages of TRL compared with those of traditional incremental conductance (INC) and five other conventional meta-heuristic algorithms. The case studies include a start-up test, step change in solar irradiation with constant temperature, stepwise change in both temperature and solar irradiation, and a daily site profile of temperature and solar irradiation in Hong Kong.

Keywords: photovoltaic systems; MPPT; partial shading condition; transfer reinforcement learning; space decomposition

1. Introduction

In the past decade, a continuous decline in the overall price of photovoltaic (PV) modules can be witnessed around the world, thanks to the advancement of new materials and manufacturing, as well as the ever-growing attention to greenhouse gas emissions [1,2]. As a consequence, solar energy has rapidly become a promising renewable power source in the global energy market. Technologically, PV systems own the elegant merits of easy installation, high safety, solar resources abundance, nearly free maintenance, and environmental friendliness [3–5]. Thus far, large-scale PV systems are widely installed, due to their short-term and long-term economic prospects [6,7].

In practice, the stochastic variation in actual environmental conditions, e.g., variation of solar radiation and fluctuation in temperature, usually leads to the power–voltage (P–V) curve to exhibit a highly nonlinear and time-varying feature. Hence, how to accurately determine the output characteristics of PV cells, as well as the maximum possible output of PV systems under various weather conditions, becomes a very challenging issue. This task is often referred to as maximum power point tracking (MPPT) [8]. For the sake of achieving MPPT, a power converter (DC–DC converter and/or inverter) is often used to connect with PV systems. Currently, conventional MPPT techniques have received further development so that, in the recent PV systems, the output power can be dynamically adjusted under different environmental conditions, e.g., hill climbing [9], perturb and observe (P&O) [10], and incremental conductance (INC) [11]. All of these schemes adopt a common assumption that the PV cells share the same module as well as the modules share the same array, and are exposed to the same temperature and solar irradiation, upon which only one maximum power point (MPP) exists. Although they own a simple structure and can efficiently seek the MPP under uniform solar irradiation conditions, a consistent oscillation around MPP is inevitable, which causes a long-lasting loss of solar energy. Besides, offline MPPT approaches such as fractional short circuit current (FSCC) [12] and fractional open circuit voltage (FOCV) [13] have been adopted for PV systems, which possess the prominent superiorities of relatively lower complexity and inexpensive implementation. Nevertheless, a common deficiency of these methods is due to the fact that they will not be applicable when solar irradiation is rapidly changing.

Furthermore, when the distribution of solar irradiation among PV modules is unequal, an uneven solar irradiation scenario may emerge, namely partial shading conditions (PSC). For example, the shadows caused by surroundings such as buildings, trees, clouds, birds, dirt, etc. Every single PV module may receive different levels of solar radiation [14]. Under this circumstance and the presence of the bypass diodes, the output P–V curve is usually nonlinear, that is, it will contain multiple local maximum power points (LMPPs) and a single global MPP (GMPP). Generally speaking, at LMPP, the PV system usually reaches a low-quality optimum point, while the aforementioned methods can be easily trapped, thus, they are inadequate to fully exploit the solar energy under PSC. To handle this intractable hindrance, a great number of approaches have been introduced. For example, reference [15] developed a fuzzy logic controller (FLC) where the approximate optimal design for membership functions and control regulations were found to be the same by GA. In addition, for the sake of achieving the rapid tracking of GMPP under PSC, a new method called the improved particle swarm optimization algorithm (PSO), based on strategy with variable sampling time, was proposed [16]. In literature [17], in order to accomplish MPPT under different environmental conditions and PSC, an artificial bee colony (ABC) algorithm was proposed, which only requires few parameters and its convergence has no relation to the initial conditions. In [18], the bio-inspired Cuckoo search algorithm (CSA) was adopted to effectively tackle PSC by the use of Levy flight with fast convergence. Moreover, a social behaviour motivated algorithm named teaching–learning-based optimization (TLBO) was adopted to achieve the accurate tracking of GMPP under PSC, the advantages of this algorithm are simple structure and fast convergence [19]. Furthermore, the generalized pattern search (GPS) optimization algorithm [20] was devised to resolve PSC, which has superior performance, such as high convergence speed, excellent dynamic, and steady state efficiencies, as well as simple operation. In reference [21], an ant colony optimization (ACO) combined with a novel strategy of pheromone updating was developed for MPPT, which can effectively improve the speed of tracking, accuracy, stability, and robustness under various weather conditions and different partial shading patterns. However, all of these meta-heuristic algorithms have two main deficiencies as they are independently utilized for MPPT under various scenarios, as follows:

- High convergence randomness: Unlike the deterministic optimization algorithms, since the meta-heuristic algorithms adopt random searching mechanisms, the final optimal solutions may be different in different runs, which will cause the output power to fluctuate greatly and is undesirable to the operation of PV systems;
- Difficult to balance the optimum quality and computation time: To obtain a high-quality optimum, the meta-heuristic algorithms usually need to establish a larger size of initial population and carry out many iterations, which results in huge computational burden and long computing time. However, considering that the MPPT's control cycle is extremely short, it is inevitable to lessen the size of population and the iteration numbers, which will lead to a significant reduction in the quality of optimization.

Rapid development of artificial intelligence in recent years, especially Google DeepMind's AlphaGo [22], which has easily defeated two world champions in two world-renowned Go matches in 2016 and 2017, respectively, has boosted a tide of artificial intelligence. In fact, the model-free reinforcement learning (RL) is one of the core algorithms of AlphaGo, which can rapidly construct an optimal action policy at each state, according to its current knowledge or experience [23]. Motivated from this outstanding characteristic, a new transfer reinforcement learning (TRL) with space decomposition for MPPT of PV systems under PSC is proposed in this paper. In comparison with the aforementioned meta-heuristic algorithms, TRL has the following two advantages:

- Capability of knowledge transfer: Through a positive knowledge transfer from past optimization tasks, the optimal knowledge matrices of the new optimization task can be approximated by TRL, hence this method can efficiently harvest an optimum of high quality;
- Capability of online learning: TRL can continuously learn new knowledge from interactions with the environment based on RL, which can rapidly adapt to MPPT under different solar irradiation, temperatures, and PSC.

2. Modelling of PV Systems under PSC

2.1. PV Cell Model

A PV cell model is usually combined in both series and parallel for the purpose of providing an output which is desired [24]. The current–voltage relationship can be given by [25,26]

$$I_{pv} = N_p I_g - N_p I_s \left(\exp\left[\frac{q}{AKT_c} \left(\frac{V_{pv}}{N_s} + \frac{R_s I_{pv}}{N_p} \right) \right] - 1 \right) \tag{1}$$

where the meaning of each symbol is given in nomenclature. Here, I_{ph} denotes the generated photocurrent that is mainly influenced by solar irradiation, which can be derived as

$$I_{ph} = (I_{sc} + k_i(T_c - T_{ref})) \frac{S}{1000} \tag{2}$$

In addition, the saturation current I_s of PV cells varies with the change of temperature on the basis of the below relationship:

$$I_s = I_{RS} \left[\frac{T_c}{T_{ref}} \right]^3 \exp\left[\frac{qE_g}{Ak} \left(\frac{1}{T_{ref}} - \frac{1}{T_c} \right) \right] \tag{3}$$

Equations (1) to (3) denote that the current produced by the PV array is simultaneously dependent on the temperature and solar irradiation.

2.2. PSC Effect

In general, the PV system needs to ensure a certain output voltage; however, a single PV cell can only output extremely low voltage (almost 0.6 V). Hence, PV cells are always connected with each other in a string to improve the output voltage. At the same time, when the array is shaded for some reason, the output voltage of the PV cells in the shaded part will be lower than that of the unshaded PV cells, due to the decline of received solar irradiation. Consequently, the shaded PV cells will consume a part of the generated power. This phenomenon causes large loss of output power in the PV string. In addition, it also leads to hot spots in the location of the shaded PV cells, which will greatly decrease the service life of PV cells [27].

To solve this issue, the shaded PV cells are usually bypassed by bypass diodes. Figure 1a demonstrates the operation in a PV array with parallel strings. Although adding bypass diodes can effectively solve the issues mentioned above in shaded PV cells, they also result in a new problem, e.g., they will distort the original P–V characteristic curves of PV cells and form a two-peak curve. In particular, such a situation turns thornier when a few PV strings are connected in parallel for the sake of obtaining a larger output current. Generally, when the number of shaded PV cells on each string changes, each string will generate various PV curves. Because of the parallel connection, those PV curves with multiple peaks are usually combined to produce a multi-peak curve illustrated in Figure 1b. Hence, in order to determine the maximum solar energy from the PV array, the PV systems ought to operate at the GMPP all the time. Only in this way, the large amount of energy will not be lost at LMPP.

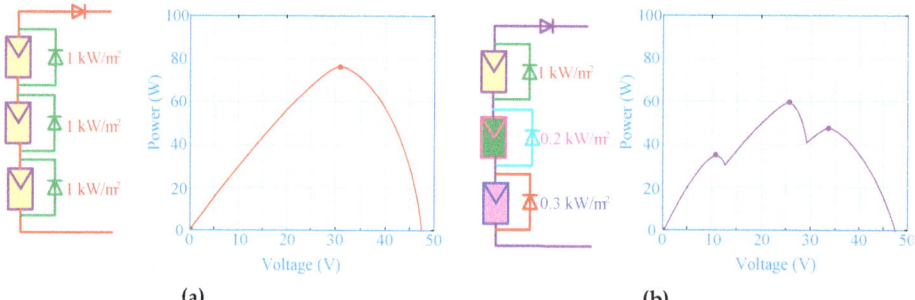

Figure 1. Partial shading conditions (PSC) effect. (**a**) Power–voltage (*P*–*V*) curve under uniform solar irradiation and temperature and (**b**) *P*–*V* curve under PSC.

3. Transfer Reinforcement Learning with Space Decomposition

The proposed TRL mainly contains two operators, i.e., the RL via uninterrupted interplay with the environment and the knowledge transfer between the previous and new tasks, as clearly illustrated in Figure 2.

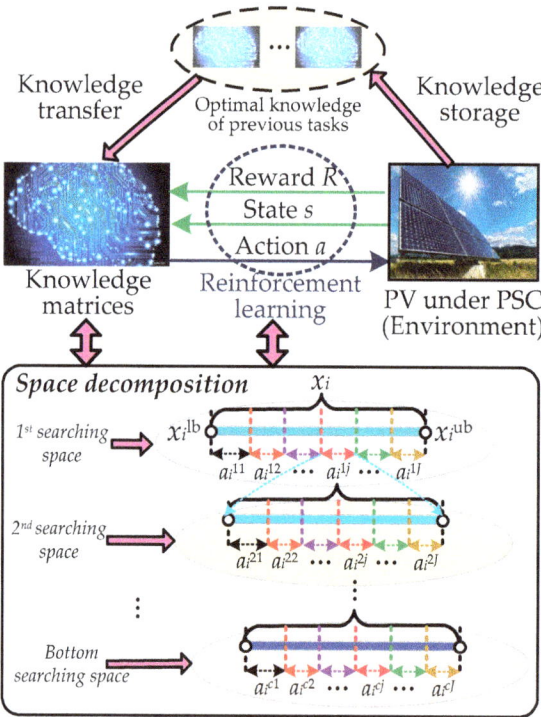

Figure 2. Principle of transfer reinforcement learning (TRL) with space decomposition.

3.1. Space Decomposition Based Reinforcement Learning

RL is a commonly used machine learning technique, which can acquire new knowledge in a dynamic environment via interaction. Here, the famous RL called Q-learning is adopted to learn the MPPT knowledge. However, if a system needs a high control accuracy, the searching space of the continuous control variable should be divided into a large number of selected actions (e.g., 10^6 actions for a continuous control variable between 0 to 1). As a result, the conventional Q-learning [28] easily encounters the curse of dimension and a low-efficiency learning rate for selecting an optimal action of a continuous control variable [29].

To handle this problem, the space decomposition is introduced to decompose the large original searching space into multi-layered smaller searching subspaces. As illustrated in Figure 3, the optimization space of the ith controllable variable x_i can be decomposed into J smaller searching subspaces in each layer. If the jth action a_i^{1j} is selected in the first layer's searching space, then the agent will seek a more accurate searching space in the corresponding second layer's searching space. Therefore, the optimization accuracy of the control variable x_i can be calculated as

$$OA_i = \frac{x_i^{ub} - x_i^{lb}}{c \cdot J} \qquad (4)$$

where c represents the number of decomposition layers; and x_i^{lb} and x_i^{ub} are the lower and upper bounds of the ith controllable variable, respectively.

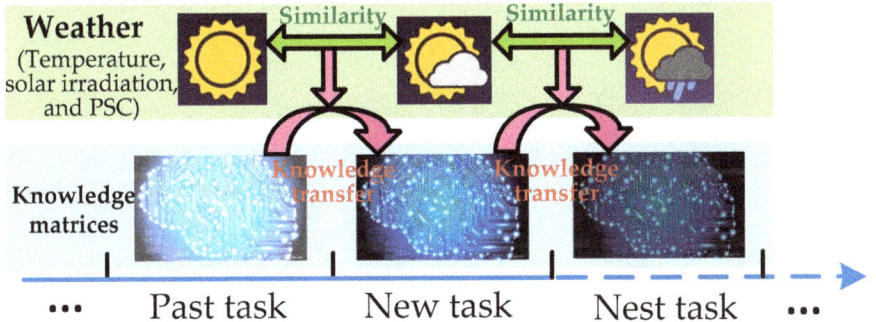

Figure 3. Knowledge transfer of TRL between two adjacent tasks for maximum power point tracking (MPPT) with PSC.

Based on Equation (4), if the number of actions in each layer is set to be 10 (i.e., $J = 10$), then the same accuracy (10^{-6}) can be achieved for a continuous control variable between 0 to 1 when $c = 6$. This means that the number of selected actions can be significantly reduced from 10^6 to 10. Therefore, the learning rate and control accuracy of Q-learning can be considerably improved, based on the space decomposition.

After selecting all the actions in all the layers, the solution of the controllable variable can be identified as

$$x_i = x_i^{c,lb} + a_i^{cj} \cdot \left(x_i^{c,ub} - x_i^{c,lb} \right) / J \qquad (5)$$

$$x_i^{l,lb} = \begin{cases} x_i^{lb}, & \text{if } l = 1 \\ x_i^{l-1,lb} + a_i^{l-1,j} \cdot \frac{\left(x_i^{l-1,ub} - x_i^{l-1,lb} \right)}{J}, & \text{otherwise} \end{cases} \qquad (6)$$

$$x_i^{l,ub} = \begin{cases} x_i^{ub}, & \text{if } l = 1 \\ x_i^{l-1,ub} + a_i^{l-1,j} \cdot \frac{\left(x_i^{l-1,ub} - x_i^{l-1,lb} \right)}{J}, & \text{otherwise} \end{cases} \qquad (7)$$

where $x_i^{l,lb}$ and $x_i^{l,ub}$ are the lower and upper bounds of the *l*th layer's searching space, respectively; while $a_i^{l,j}$ is the *j*th action in the *l*th layer's searching space.

3.2. Knowledge Update

According to the learning mechanism of Q-learning, the knowledge matrix can be updated based on the executed state–action pair with the feedback reward. By combining the space decomposition, the knowledge matrix of each searching space layer can be updated as [28]:

$$Q_{i,k}^l\left(s_{i,k}^l, a_{i,k}^l\right) = Q_{i,k}^l\left(s_{i,k}^l, a_{i,k}^l\right) \\ + \alpha \left[R_{i,k}^l\left(s_{i,k}^l, s_{i,k+1}^l, a_{i,k}^l\right) + \gamma \max_{a \in A_i^l} Q_{i,k}^l\left(s_{i,k+1}^l, a\right) - Q_{i,k}^l\left(s_{i,k}^l, a_{i,k}^l\right) \right] \quad (8)$$

where Q_i^l represents the knowledge matrix of the *l*th layer' searching space for the *i*th controllable variable; $\left(s_{i,k}^l, a_{i,k}^l\right)$ is the state–action pair executed at the *k*th iteration, with $k = 1, 2, \ldots, k_{max}$; k_{max} represents the maximum iteration number; α is the knowledge learning factor, with $\alpha \in (0, 1)$; γ denotes the discount factor, with $\gamma \in (0, 1)$; R_i^l is the reward function; and A_i^l means the action space of the *l*th layer's searching space, respectively.

It can be seen from Equation (8) that at each iteration, only one element of each knowledge matrix can be updated since the conventional Q-learning employs a single RL agent for exploration and exploitation in a dynamic environment. Consequently, it will lead to a slow learning rate; thus a high-quality optimal solution cannot be rapidly obtained for a real-time control of PV systems. Hence, a cooperative swarm is employed to further accelerate the learning rate, as it can simultaneously update multiple elements of each knowledge matrix with multiple state-action pairs. Similar to (8), each knowledge matrix of TRL can be updated by [30]

$$Q_{i,k}^l\left(s_{i,k}^{l,m}, a_{i,k}^{l,m}\right) = Q_{i,k}^l\left(s_{i,k}^{l,m}, a_{i,k}^{l,m}\right) \\ + \alpha \left[R_{i,k}^{l,m}\left(s_{i,k}^{l,m}, s_{i,k+1}^{l,m}, a_{i,k}^{l,m}\right) + \gamma \max_{a \in A_i^l} Q_{i,k}^l\left(s_{i,k+1}^{l,m}, a\right) \\ - Q_{i,k}^l\left(s_{i,k}^{l,m}, a_{i,k}^{l,m}\right) \right], \quad m = 1, 2, \ldots, M. \quad (9)$$

where *M* represents the population size of the cooperative swarm.

3.3. Exploration and Exploitation

In general, a wide exploration will enhance the possibility of searching a global optimum, but will also consume additional computation time. In contrast, a deep exploitation will enhance the convergence speed, but will easily result in a local optimum in low quality. In order to keep exploitation and exploration in balance, the ε-Greedy rule [31] is adopted to select actions on the basis of the current knowledge matrices, which yields

$$a_{i,k+1}^{l,m} = \begin{cases} \arg\max_{a_i^l \in A_i^l} Q_{i,k}^l\left(s_{i,k+1}^{l,m}, a_i^l\right), & \text{if } q_0 < \varepsilon \\ a_{rand}, & \text{otherwise} \end{cases} \quad (10)$$

where q_0 is a uniform random number between 0 and 1; ε is the rate of exploitation, i.e., the possibility of selecting the greedy action; and a_{rand} represents a stochastic action in the action space, i.e., the global search for avoiding a low-quality local optimum, respectively.

3.4. Knowledge Transfer

Through exploiting the optimal knowledge matrices of the previous tasks, the knowledge transfer [32] can approximate the optimal knowledge matrices of a new task, and this is how knowledge transfer works. In this study, the most similar previous task will be chosen for knowledge transfer, based on its similarity with the new task, which can be expressed as

$$Q_i^{n0} = r \cdot Q_i^{s*} + (1-r) \cdot Q_i^{initial} \tag{11}$$

where Q_i^{n0} is the approximated optimal knowledge matrices of the ith controllable variable of the new task; Q_i^{s*} denotes the optimal knowledge matrices of the ith controllable variable of the most similar previous task; $Q_i^{initial}$ represents the initial knowledge matrices of the new task without knowledge transfer; and r represents the comparability between the most similar previous task and the updated task, with $0 \le r \le 1$, respectively.

4. TRL Design of PV Systems for MPPT

4.1. Control Variable and Action Space

For the purpose of obtaining the GMPP of a PV system, the output voltage V_{pv} is chosen as the control variable, in which the entire searching space is decomposed into four layers. In each layer, the searching space is uniformly discretized into ten actions within the corresponding range from lower bounds to upper bounds.

4.2. Reward Function

For a given output voltage V_{pv}, the PV system can generate the corresponding power under the current solar irradiation, temperature, and PSC. In TRL, the higher the quality of the solution is, the larger reward the individual will receive. Based on this rule, the reward function can be designed as [30]:

$$R_{i,k}^{l,m}\left(s_{i,k}^{l,m}, s_{i,k+1}^{l,m}, a_{i,k}^{l,m}\right) = \begin{cases} \max_{m=1,2,\ldots,M} f\left(V_{pv}^m\right), & \text{if } \left(s_{i,k}^{l,m}, a_{i,k}^{l,m}\right) \in SA_k^{best} \\ 0, & \text{otherwise} \end{cases} \tag{12}$$

where V_{pv}^m is the obtained solution by the mth individual and SA_k^{best} denotes the explored state–action pairs set of the best individual with the maximum power output at the kth iteration.

4.3. Knowledge Transfer

It is clear that the aforementioned three conditions, e.g., solar irradiation, temperature, and PSC, can be considered as the main similarities between various optimization tasks. On the other hand, the similarity between two adjacent optimization tasks is usually very high, since these weather conditions cannot vary dramatically in a very short time. Hence, the optimal knowledge matrices of the adjacent past task is chosen for knowledge transfer to the new task (See Figure 3), while the similarity described in (11) can be designed as

$$r = 1 - \frac{|T_c^n - T_c^p|}{T_{ref}} - \prod_{w=1}^{N_s \cdot N_p} \frac{|S_w^n - S_w^p|}{S_{ref}} \tag{13}$$

where T_c^n and T_c^p are the temperatures of the new task and the past task, respectively;

4.4. Overall Execution Procedure

For the PV system, the overall flow diagram of TRL to achieve MPPT under PSC is illustrated in Figure 4. Firstly, the original searching space of output voltage is decomposed into a four-layered smaller searching subspace within its corresponding lower bounds and upper bounds. Then, the knowledge transfer between the new task and the past task is implemented according to their similarity of weather conditions. Furthermore, TRL can update the knowledge matrices via multiple explorations and exploitations in the scheduled iterations. At last, for the PV system, the optimal solution (optimal output voltage) can be obtained to achieve MPPT under PSC.

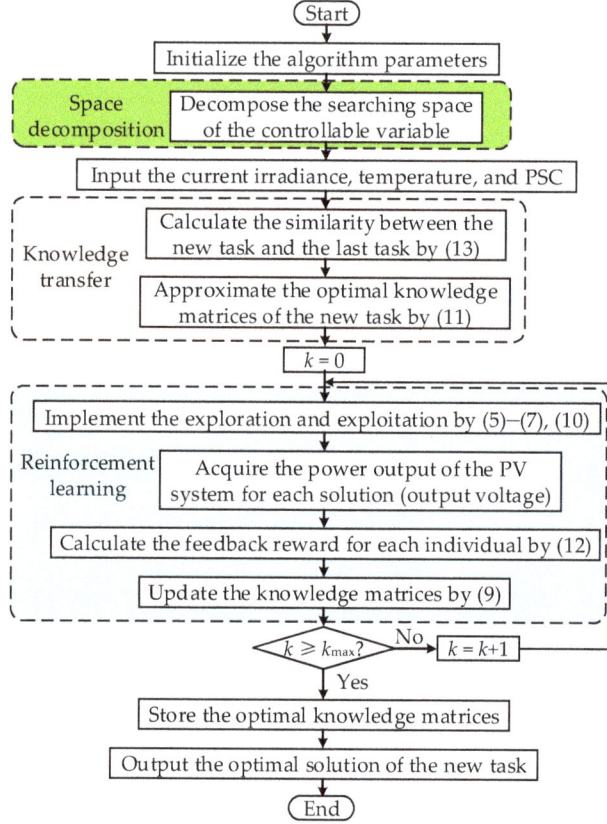

Figure 4. Overall flow diagram of TRL for MPPT. PSC: Partial Shading Conditions.

5. Case Studies

To further analyze the MPPT practicability of TRL under PSC, it was compared with that of INC [11], GA [15], PSO [16], ABC [17], CSA [18], and TLBO [19], respectively. Four case studies are carried out in this section. Here, each meta-heuristic algorithm shares the same optimization cycle, which is chosen as 0.01 s. Meanwhile, the TRL parameters are given in Table 1.

Table 1. The parameters of TRL. TRL: Transfer Reinforcement Learning.

Parameter	Range	Value
J	$J > 1$	10
c	$c > 1$	4
α	$0 < \alpha < 1$	0.01
γ	$0 < \gamma < 1$	0.0001
ϵ	$0 < \epsilon < 1$	0.9
k_{max}	$k_{max} > 1$	5
M	$M > 1$	5

For MPPT under PSC, a buck–boost converter is employed, due to its advantages described in reference [33]. Table 2 demonstrates the parameters of the PV system. In addition, the rated values of environment temperature and solar irradiation are set as 25 °C and 1000 W/m², respectively.

Table 2. The photovoltaic (PV) system parameters.

Typical peak power	51.716 W	Nominal operation cell temperature (T_{ref})	25 °C
Voltage at peak power	18.47 V	Factor of PV technology (A)	1.5
Current at peak power	2.8 A	Switching frequency (f)	100 kHz
Short-circuit current (I_{sc})	1.5 A	Inductor (L)	500 mH
Open-circuit voltage (V_{oc})	23.36 V	Resistive load (R)	200 Ω
Temperature coefficient of I_{sc} (k_1)	3 mA/°C	Capacitor (C_1, C_2)	1 µF

5.1. Start-Up Test

The first step to simulate the PSC is to set the solar irradiation of three PV strings to be 200 W/m², 300 W/m², and 1000 W/m², respectively. The online optimization responses of various methods for MPPT are illustrated in Figure 5. It is clear that INC can easily reach the point of steady convergence in far less time than the other methods. However, it has a vital drawback in that it cannot make an effective distinction between GMPP and LMPP, which means it might often be trapped at a low-quality local optimum as it is readily stagnated at an MPP. Generally speaking, due to their significant ability of global searching, other meta-heuristic algorithms can usually find a better quality optimum with larger power and energy. Among them, TRL owns the highest convergence stability as it can avoid a blind/random search by the use of knowledge transfer.

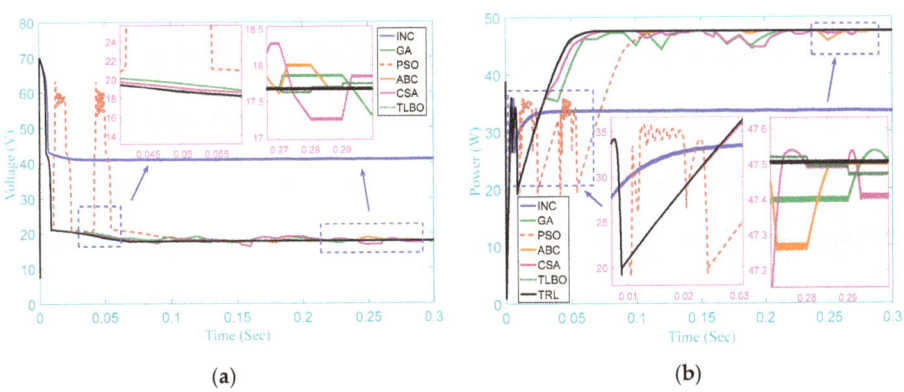

Figure 5. PV system responses of seven methods obtained on the start-up test. (**a**) Voltage; (**b**) Power.

5.2. Step Change in Solar Irradiation with Constant Temperature

As shown in Figure 6, the core process is to impose a set of solar irradiation steps on the PV array, where the step change is applied every second. The temperature is maintained to be constant at 25 °C during the whole test. The online optimization outcomes of various approaches for MPPT with step change solar irradiations are illustrated in Figure 7. It can be found that the obtained results are similar to those of the start-up test. The output power and voltage derived by those meta-heuristic algorithms, except TRL, are relatively prone to volatility if the solar irradiation is not always steady and varies at a dramatic pace. This also verifies that the knowledge transfer can effectively guarantee the convergence stability of TRL, i.e., the control strategies of adjacent optimization tasks only have a slight difference under the same weather conditions.

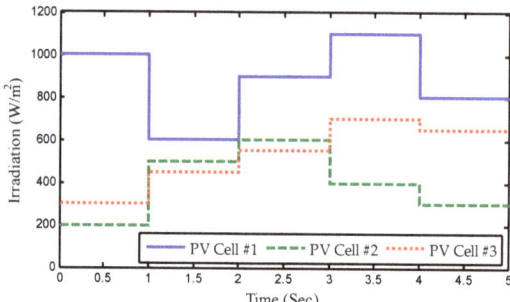

Figure 6. Step change of solar irradiation with PSC. PSC: Partial Shading Conditions.

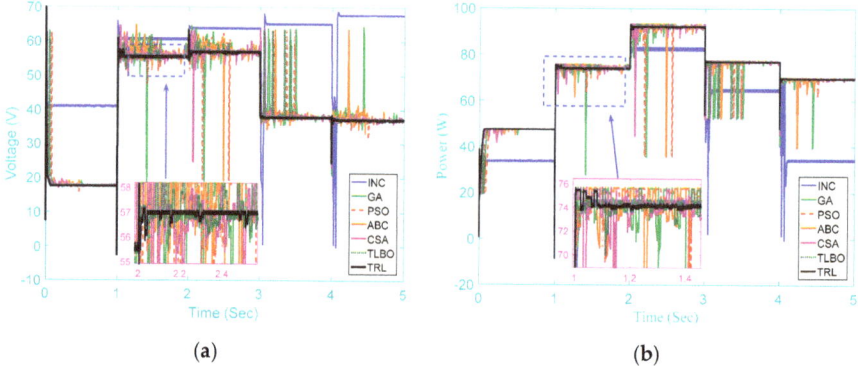

Figure 7. PV system responses of seven methods obtained on the step change in solar irradiation with constant temperature. (**a**) Voltage; (**b**) Power.

5.3. Gradual Change in Both Solar Irradiation and Temperature

Figures 8 and 9 show the procured results of seven algorithms for MPPT when solar irradiation and temperature both change gradually. A conclusion can be drawn that, except for TRL, the other meta-heuristic algorithms are still prone to generating the larger power fluctuations, even when the solar irradiation and temperature change slowly. Due to the beneficial guidance by knowledge transfer, TRL can significantly alleviate the power fluctuations without a blind/random search.

This also reveals that, for real-time MPPT, TRL is capable of speedily seeking an optimum of high quality through the space decomposition on the basis of RL and beneficial knowledge transfer.

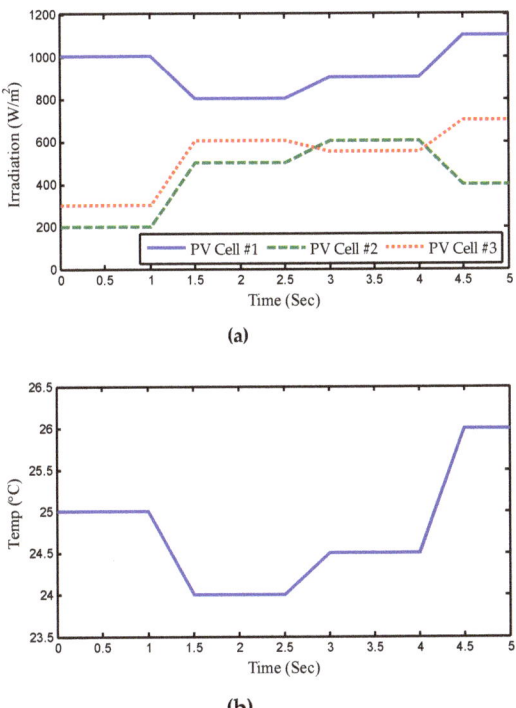

Figure 8. Gradual change in both solar irradiation and temperature. (**a**) Irradiation and (**b**) temperature.

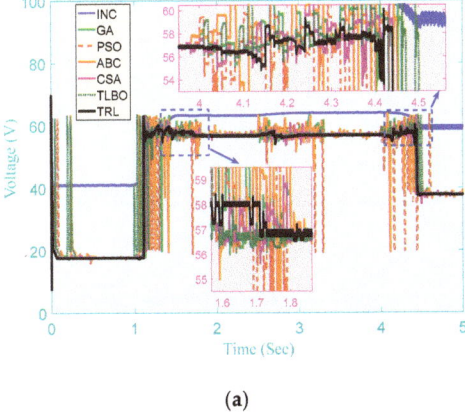

Figure 9. *Cont.*

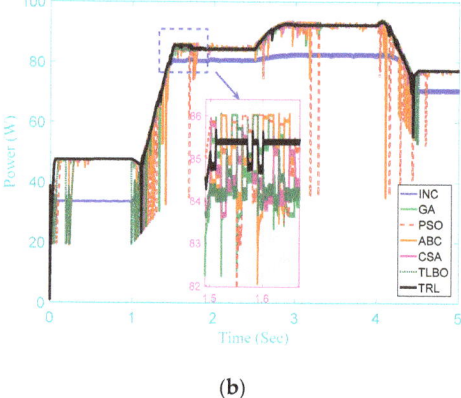

(b)

Figure 9. PV system responses of seven methods obtained on the gradual change in both solar irradiation and temperature. (**a**) Voltage; (**b**) Power.

5.4. Daily Field Profile of Solar Irradiation and Temperature in Hong Kong

For the purpose of testing the specific practicability of TRL in practical application, the temperature and solar irradiation measured in Hong Kong was used to simulate the PV system for MPPT (See Figures 10 and 11). The metrical data are mainly selected from four representative days of four different seasons in 2016, in which the interval of data is set to 10 min. Note that the randomness and intermittence of solar energy and renewable energy system (RES) [34–37] is a very common issue usually resulting from uncertain atmospheric conditions.

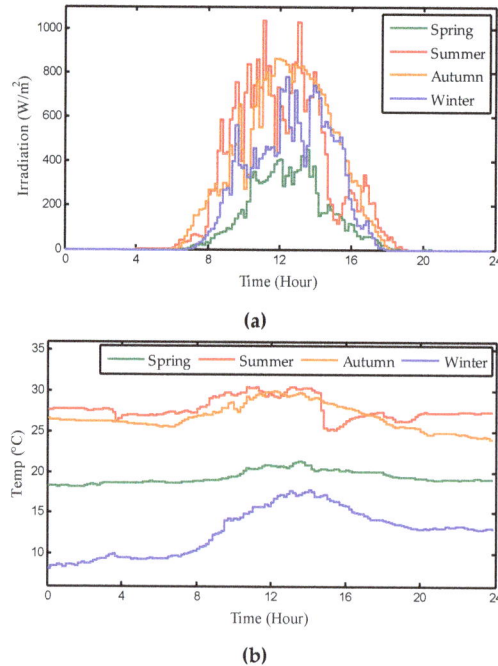

Figure 10. Daily profile of solar irradiation and temperature in Hong Kong. (**a**) Irradiation; (**b**) Temperature.

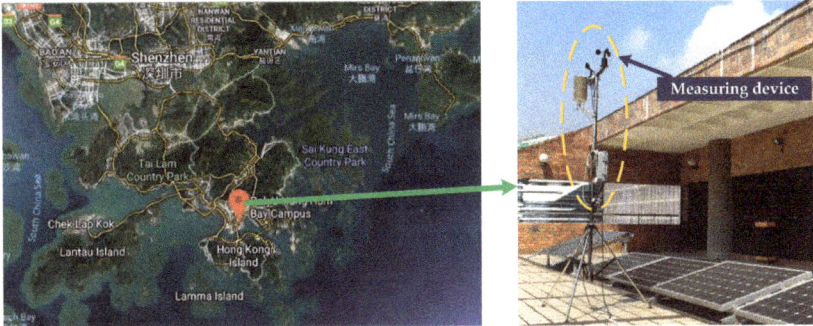

Figure 11. The detailed geographical position of the measuring device for solar irradiation and temperature.

Figures 12 and 13 demonstrate the output power of seven algorithms for MPPT in different seasons. It can be well illustrated that, compared with INC, in the PV system, all the meta-heuristic algorithms can obtain more output power, where the output energy of TRL reaches 115.52% of that of INC in the spring. That aside, one can derive that although the performances of all meta-heuristic algorithms are comparatively small during the whole simulation period, TRL can still outperform other algorithms, which means that it can always give out the most power in any season.

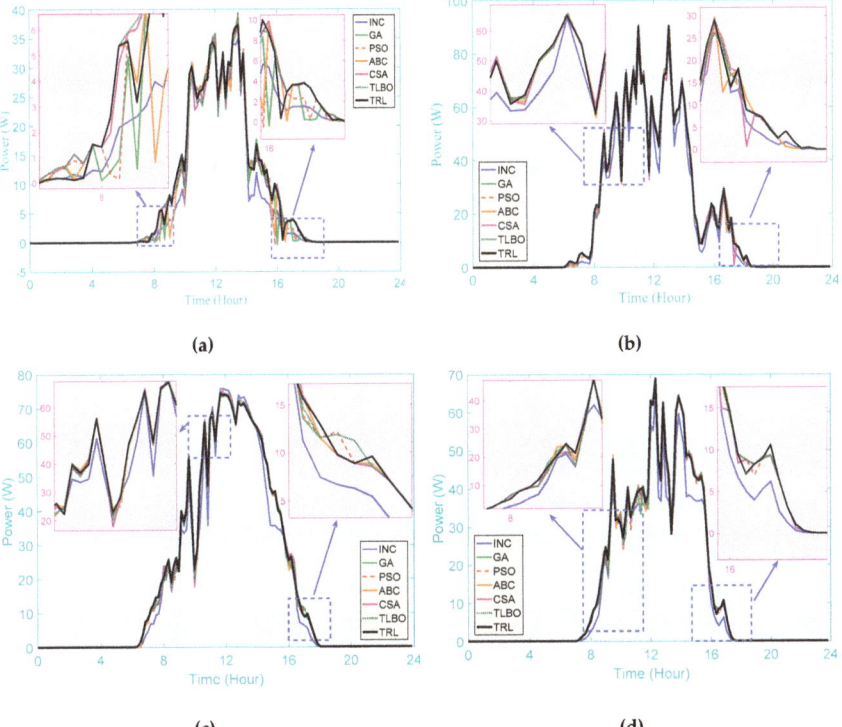

Figure 12. PV system responses obtained on a typical day in Hong Kong. (**a**) Spring; (**b**) summer; (**c**) autumn; (**d**) winter.

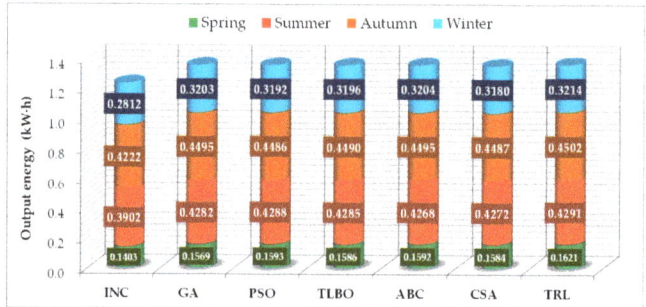

Figure 13. Statistical results of output energy of the PV system obtained by seven algorithms in different seasons.

6. Conclusions

A novel method called TRL using space decomposition has been proposed in this paper, which is designed for PV systems to obtain the maximum attainable solar energy under PSC, whose contributions can be summarized as follows:

(1) Through space decomposition, TRL can efficiently learn the knowledge for MPPT with PSC in real time; thus a high-quality optimum can be obtained to ensure that the PV system produces more energy under various environmental conditions;
(2) The knowledge transfer can effectively avoid a blind/random search and provide a beneficial guidance to TRL, which results in a fast convergence and a high convergence stability. Therefore, not only can the output power be maximized for the PV system under various scenarios, but the power fluctuation can also be significantly reduced as the weather condition varies;
(3) Compared with the conventional INC and other typical meta-heuristic algorithms, the TRL-based MPPT algorithm can produce the largest amount of output energy in the presence of PSC and other time-varying atmospheric conditions, which can bring about considerable economic benefit for operation in the long term.

Author Contributions: Preparation of the manuscript was performed by M.D., D.L., C.Y., S.L., Q.F., B.Y., and X.Z.

Acknowledgments: The authors gratefully acknowledge the support of ADN Comprehensive Demonstration Project of Smart Grid Application Demonstration Area in Suzhou Industrial Park and Yunnan Provincial Basic Research Project—Youth Researcher Program (2018FD036).

Conflicts of Interest: The authors declare no conflict of interest.

References

1. Yang, B.; Yu, T.; Zhang, X.S.; Li, H.F.; Shu, H.C.; Sang, Y.Y.; Jiang, L. Dynamic leader based collective intelligence for maximum power point tracking of PV systems affected by partial shading condition. *Energy Convers. Manag.* **2019**, *179*, 286–303. [CrossRef]
2. Wang, X.T.; Barnett, A. The evolving value of photovoltaic module efficiency. *Appl. Sci.* **2019**, *9*, 1227. [CrossRef]
3. Yang, B.; Yu, T.; Shu, H.C.; Dong, J.; Jiang, L. Robust sliding-mode control of wind energy conversion systems for optimal power extraction via nonlinear perturbation observers. *Appl. Energy* **2018**, *210*, 711–723. [CrossRef]
4. Worighi, I.; Geury, T.; Baghdadi, M.E.; Mierlo, J.V.; Hegazy, O.; Maach, A. Optimal design of hybrid PV-Battery system in residential buildings: End-user economics, and PV penetration. *Appl. Sci.* **2019**, *9*, 1022. [CrossRef]
5. Hyunkyung, S.; Zong, W.G. Optimal design of a residential photovoltaic renewable system in South Korea. *Appl. Sci.* **2019**, *9*, 1138.

6. Yang, B.; Zhong, L.E.; Yu, T.; Li, H.F.; Zhang, X.S.; Shu, H.C.; Sang, Y.Y.; Jiang, L. Novel bio-inspired memetic salp swarm algorithm and application to MPPT for PV systems considering partial shading condition. *J. Clean. Prod.* **2019**, *215*, 1203–1222. [CrossRef]
7. Gulkowski, S.; Zdyb, A.; Dragon, P. Experimental efficiency analysis of a photovoltaic system with different module technologies under temperate climate conditions. *Appl. Sci.* **2019**, *9*, 141. [CrossRef]
8. Wu, Z.; Yu, D. Application of improved bat algorithm for solar PV maximum power point tracking under partially shaded condition. *Appl. Soft Comput.* **2018**, *62*, 101–109. [CrossRef]
9. Tanaka, T.; Toumiya, T.; Suzuki, T. Output control by hill-climbing method for a small scale wind power generating system. *Renew. Energy* **2014**, *12*, 387–400. [CrossRef]
10. Mohanty, S.; Subudhi, B.; Ray, P.K. A grey wolf-assisted Perturb & Observe MPPT algorithm for a PV system. *IEEE Trans. Energy Convers.* **2017**, *32*, 340–347.
11. Zakzouk, N.E.; Elsaharty, M.A.; Abdelsalam, A.K.; Helal, A.A. Improved performance low-cost incremental conductance PV MPPT technique. *IET Renew. Power Gener.* **2016**, *10*, 561–574. [CrossRef]
12. Sher, H.A.; Murtaza, A.F.; Noman, A.; Addoweesh, K.E.; Al-Haddad, K.; Chiaberge, M. A new sensorless hybrid MPPT algorithm based on fractional short-circuit current measurement and P&O MPPT. *IEEE Trans. Sustain. Energy* **2015**, *6*, 1426–1434.
13. Huang, Y.P. A rapid maximum power measurement system for high-concentration Photovoltaic modules using the fractional open-circuit voltage technique and controllable electronic load. *IEEE J. Photovolt.* **2014**, *4*, 1610–1617. [CrossRef]
14. Rezk, H.; Fathy, A.; Abdelaziz, A.Y. A comparison of different global MPPT techniques based on meta-heuristic algorithms for photovoltaic system subjected to partial shading conditions. *Renew. Sustain. Energy Rev.* **2017**, *74*, 377–386. [CrossRef]
15. Messai, A.; Mellit, A.; Guessoum, A.; Kalogirou, S.A. Maximum power point tracking using a GA optimized fuzzy logic controller and its FPGA implementation. *Sol. Energy* **2011**, *85*, 265–277. [CrossRef]
16. Babu, T.S.; Rajasekar, N.; Sangeetha, K. Modified Particle Swarm Optimization technique based Maximum Power Point Tracking for uniform and under partial shading condition. *Appl. Soft Comput.* **2015**, *34*, 613–624. [CrossRef]
17. Benyoucef, A.S.; Chouder, A.; Kara, K.; Silvestre, S.; Sahed, O.A. Artificial bee colony based algorithm for maximum power point tracking (MPPT) for PV systems operating under partial shaded conditions. *Appl. Soft Comput.* **2015**, *32*, 38–48. [CrossRef]
18. Ahmed, J.; Salam, Z. A maximum power point tracking (MPPT) for PV system using Cuckoo search with partial shading capability. *Appl. Energy* **2014**, *119*, 118–130. [CrossRef]
19. Rezk, H.; Fathy, A. Simulation of global MPPT based on teaching-learning-based optimization technique for partially shaded PV system. *Electr. Eng.* **2017**, *99*, 847–859. [CrossRef]
20. Javed, M.Y.; Murtaza, A.F.; Ling, Q.; Qamar, S.; Gulzar, M.M. A novel MPPT design using generalized pattern search for partial shading. *Energy Build.* **2016**, *133*, 59–69. [CrossRef]
21. Titri, S.; Larbes, C.; Toumi, K.Y.; Benatchba, K. A new MPPT controller based on the Ant colony optimization algorithm for Photovoltaic systems under partial shading conditions. *Appl. Soft Comput.* **2017**, *58*, 465–479. [CrossRef]
22. Silver, D.; Huang, A.; Maddison, C.J.; Guez, A.; Sifre, L.; Van Den Driessche, G.; Dieleman, S. Mastering the game of Go with deep neural networks and tree search. *Nature* **2016**, *529*, 484–489. [CrossRef] [PubMed]
23. Sutton, R.S.; Barto, A.G. *Reinforcement Learning: An Introduction*; MIT Press: Cambridge, MA, USA, 1998.
24. Qi, J.; Zhang, Y.; Chen, Y. Modeling and maximum power point tracking (MPPT) method for PV array under partial shade conditions. *Renew. Energy* **2014**, *66*, 337–345. [CrossRef]
25. Lalili, D.; Mellit, A.; Lourci, N.; Medjahed, B.; Berkouk, E.M. Input output feedback linearization control and variable step size MPPT algorithm of a grid-connected photovoltaic inverter. *Renew. Energy* **2011**, *36*, 3282–3291. [CrossRef]
26. Lalili, D.; Mellit, A.; Lourci, N.; Medjahed, B.; Boubakir, C. State feedback control and variable step size MPPT algorithm of three-level grid-connected photovoltaic inverter. *Sol. Energy* **2013**, *98*, 561–571. [CrossRef]
27. Chen, K.; Tian, S.; Cheng, Y.; Bai, L. An improved MPPT controller for photovoltaic system under partial shading condition. *IEEE Trans. Sustain. Energy* **2017**, *5*, 978–985. [CrossRef]
28. Watkins, J.C.H.; Dayan, P. Q-learning. *Mach. Learn.* **1992**, *8*, 279–292. [CrossRef]

29. Er, M.J.; Deng, C. Online tuning of fuzzy inference systems using dynamic fuzzy Q-learning. *IEEE Trans. Syst. Man Cybern. Part B* **2004**, *34*, 1478–1489. [CrossRef]
30. Zhang, X.; Yu, T.; Yang, B.; Cheng, L. Accelerating bio-inspired optimizer with transfer reinforcement learning for reactive power optimization. *Knowl. Based Syst.* **2017**, *116*, 26–38. [CrossRef]
31. Bianchi, R.A.C.; Celiberto, L.A.; Santos, P.E.; Matsuura, J.P.; Lopez de Mantaras, R. Transferring knowledge as heuristics in reinforcement learning: A case-based approach. *Artif. Intell.* **2015**, *226*, 102–121. [CrossRef]
32. Pan, J.; Wang, X.; Cheng, Y.; Cao, G. Multi-source transfer ELM-based Q learning. *Neurocomputing* **2014**, *137*, 57–64. [CrossRef]
33. Ishaque, K.; Salam, Z.; Amjad, M.; Mekhilef, S. An improved particle swarm optimization (PSO)-based MPPT for PV with reduced steady-state oscillation. *IEEE Trans. Power Electron.* **2012**, *27*, 3627–3638. [CrossRef]
34. Li, G.D.; Li, G.Y.; Zhou, M. Model and application of renewable energy accommodation capacity calculation considering utilization level of interprovincial tie-line. *Prot. Control Mod. Power Syst.* **2019**, *4*, 1–12. [CrossRef]
35. Tummala, S.L.V.A. Modified vector controlled DFIG wind energy system based on barrier function adaptive sliding mode control. *Prot. Control Mod. Power Syst.* **2019**, *4*, 34–41.
36. Faisal, R.B.; Purnima, D.; Subrata, K.S.; Sajal, K.D. A survey on control issues in renewable energy integration and microgrid. *Prot. Control Mod. Power Syst.* **2019**, *4*, 87–113.
37. Dash, P.K.; Patnaik, R.K.; Mishra, S.P. Adaptive fractional integral terminal sliding mode power control of UPFC in DFIG wind farm penetrated multimachine power system. *Prot. Control Mod. Power Syst.* **2018**, *3*, 79–92. [CrossRef]

© 2019 by the authors. Licensee MDPI, Basel, Switzerland. This article is an open access article distributed under the terms and conditions of the Creative Commons Attribution (CC BY) license (http://creativecommons.org/licenses/by/4.0/).

MDPI\
St. Alban-Anlage 66\
4052 Basel\
Switzerland\
Tel. +41 61 683 77 34\
Fax +41 61 302 89 18\
www.mdpi.com

Applied Sciences Editorial Office\
E-mail: applsci@mdpi.com\
www.mdpi.com/journal/applsci

www.ingramcontent.com/pod-product-compliance
Lightning Source LLC
LaVergne TN
LVHW070705100526
838202LV00013B/1034